Synthesis Lectures on Human-Centered Informatics

Series Editor

John M. Carroll, College of Information Sciences and Technology, Penn State University, University Park, PA, USA

This series publishes short books on Human-Centered Informatics (HCI), at the intersection of the cultural, the social, the cognitive, and the aesthetic with computing and information technology. Lectures encompass a huge range of issues, theories, technologies, designs, tools, environments, and human experiences in knowledge, work, recreation, and leisure activity, teaching and learning, etc. The series publishes state-of-the-art syntheses, case studies, and tutorials in key areas. It shares the focus of leading international conferences in HCI.

Sergio Sayago

Cultures
in Human-Computer
Interaction

 Springer

Sergio Sayago
University of Lleida (Campus Igualada)
Igualada, Spain

ISSN 1946-7680　　　　　　　ISSN 1946-7699　(electronic)
Synthesis Lectures on Human-Centered Informatics
ISBN 978-3-031-30245-9　　　　ISBN 978-3-031-30243-5　(eBook)
https://doi.org/10.1007/978-3-031-30243-5

This Springer imprint is published by the registered company Springer Nature Switzerland AG
The registered company address is: Gewerbestrasse 11, 6330 Cham, Switzerland

Acknowledgments

I would like to thank John M. Caroll for providing me with the opportunity to publish this book in this Synthesis Lectures series. I would also like to thank Christine Kiilerich and Prasanna Kumar Narayanasamy from Springer for their editorial assistance and patience with me. I also thank the reviewers. I am still impressed by the detailed, inspiring, and thoughtful comments and suggestions. I am indebted to Paula Forbes for her unconditional support. I feel very lucky to have Paula as a colleague and, most importantly, as a friend. I also thank those who have believed in and supported me throughout all these years in my academic career. My special thanks to Josep Blat (Universitat Pompeu Fabra), I have learned much from and with him.

Contents

About the Author

Sergio Sayago is a Lecturer (Assistant Professor) in Interactive Systems and Languages at the University of Lleida. Prior to that, he was a post-doctoral researcher at the University of Dundee (Scotland, 2010–2012, *Beatriu de Pinós* fellowship) and Universidad Carlos III de Madrid (Spain, 2012–2014, *Alliance 4 Universities* fellowship), and a visiting lecturer at the University of Lleida (Catalonia, 2014–2016) and University of Barcelona (2016–2019). Sergio is an HCI scholar interested in the human (ageing) side of digital technologies. His previous research includes web accessibility, usability, and eLearning. Since 2009, his research focuses on ageing, older people, and technology. He is the author of more than 50 publications. He won the Best Technical Paper Award in ACM W4A in 2009. He is an Associate Editor of the *International Journal of Human-Computer Studies*. Sergio has served as Associate Chair in the Aging and Accessibility subcommittee of ACM CHI in 2022 and 2023. Sergio holds a Ph.D. *cum laude* in Computer Science and Human-Computer Interaction from Universitat Pompeu Fabra (Barcelona, 2009).

Introduction

1

This introductory chapter presents the motivations for this synthesis and its objectives. It also outlines the overall perspective on culture that is adopted herein and the contributions this publication makes to Human–Computer Interaction (HCI). This chapter also provides an overview of the structure and contents of the synthesis.

1.1 Motivation, Objectives, and Readership

There are three main motivations for this book. Firstly, a great deal of HCI research in which culture is a significant theme is scattered through conference proceedings, journal publications, and books. These publications tend to focus on particular subjects, such as cross-cultural communication, participatory design, usability, and user interface design. This situation makes it difficult for HCI scholars to develop an overall view of HCI research on (and with) culture. Secondly, leading HCI textbooks, which play a key role in the development and dissemination of the field, present interesting yet partial views of culture. For example, *Interaction Design: beyond Human–Computer Interaction* (Sharp et al., 2019), *Designing the User Interface: Strategies for Effective Human–Computer Interaction* (Shneiderman et al., 2018) and *Human–Computer Interaction* (Dix et al., 2004) tend to concentrate almost exclusively on user interface elements, such as the comprehensibility of icons and metaphors across cultures (Pappachan & Ziefle, 2008; Russo & Boor, 1993), and cultural probes (Gaver et al., 1999), which attract growing research attention (Almohamed & Vyas, 2019; Brianza et al., 2022; Rodríguez et al., 2020). Other relevant aspects of culture within HCI are mostly overlooked, such as the different approaches adopted to examine this subject (e.g., quantitative, qualitative, or a combination of them), the main functions of culture in HCI (e.g., meaning, explaining differences and reducing complexity), the challenges of working with people with different cultures than that of

S. Sayago, *Cultures in Human-Computer Interaction*, Synthesis Lectures
on Human-Centered Informatics, https://doi.org/10.1007/978-3-031-30243-5_1

the researchers (Kamppuri, 2012; Sabie et al., 2022), or how we actually design usable and meaningful technologies for culturally diverse users (Aykin, 2005; Singh & Pereira, 2005; Sun, 2012). Thirdly, the HCI community has developed an increased but moderate interest in culture. The prevailing methodology is comparative, experimental, quantitative and uses national groups as cultural groups (Clemmensen & Roese, 2010; Kamppuri et al., 2006; Linxen et al., 2021). As discussed in this book, culture is a complex and key dimension in HCI, and examining it requires a more eclectic approach (e.g., combining different methodologies and going beyond the notion of culture = nation).

This book aims to (1) provide a synthesis of the topic of culture in the context of HCI and (2) give a current overview of HCI research in which culture is a primary topic of interest. By doing so, this book aspires to contribute to HCI with a richer understanding of culture. The research literature this book draws on is wide-ranging, including Anthropology, Sociology, Cultural Studies, Robotics, Disability Studies, Cultural and Cross-Cultural Psychology, Usability, and Design, in addition to HCI. This includes over 40 books, and 400 conference and journal papers published in several academic databases (ACM Digital Library, SCOPUS, Springer, IEEE Xplore, and Web of Science). The bibliography at the end of the book shows a great deal of the literature examined for the preparation of this synthesis. Representative studies of this literature and, in some cases, exemplary ones are cited throughout this book. Given that a synthesis is not only expected to bring together an important body of research but also generate ideas for future research, another objective of this book is to discuss some open issues in an attempt to spark further investigation and discussion.

This book does not aim to be the definitive synthesis of culture in HCI research. This book is partial and limited—as any publication dealing with culture probably is. Yet, this synthesis aspires to become a useful overview of culture for HCI scholars with different levels of academic seniority. Senior HCI scholars can find herein a structured and integrated summary of works that might help them develop an overall picture of HCI research on culture and push their investigations forward. For HCI scholars with a lower level of seniority (e.g., Ph.D. students), this book provides them with a guided and short introduction to a large body of research on a complex issue. This book is written for an HCI audience.

1.2   Culture in This Book

There are many different definitions of culture (Baldwin et al., 2006). Thus, before moving on, it might be important to clarify what culture means herein. Doing so runs the risk of inevitably reducing the term culture to the concrete and specific, which is a way of valorizing some definitions and ignoring or overlooking others (Tredinnick, 2008). This book provides several definitions of culture across disciplines. It also summarizes the main views of culture within HCI (e.g., "software of the mind" (Hofstede et al., 2010) and culture as communication (Hall, 1973, 1976, 1982). Yet, it is necessary to discuss the overall perspective or stance on culture that is adopted, rather than talking about culture in general (i.e., without any definition, direction, or purpose), to help the reader understand which aspects are (not) considered.

Overall, culture can be regarded as either a definitive or sensitizing concept.[1] A definitive concept of culture provides a clear definition of culture. For example, culture "is the collective programming of the mind that distinguishes the members of one group or category of people from others" (Hofstede et al., 2010, p. 6). A sensitizing concept of culture encourages us to consider that what we are referring to the concept of culture depends on each empirical instance. Sensitizing concepts provide some direction and purpose (Hertzum, 2018; Puddephatt, et al., 2009, p. 19). This book sees culture from a sensitizing point of view. Cultures (in the plural) in the book's title is not a typo. People fill the term culture with meaning depending on their academic background and also on their personal experience. Music, literature, lifestyle, power, identity, rebellion, food, and meaning are, to name a few, several ways of understanding culture. People might also belong simultaneously to multiple cultures regardless of their national one (English-Lueck, 2017; Epp et al., 2022; Tukiainen, 2010). This book acknowledges and embraces this diversity, which emanates from HCI research on (and with) culture. Yet, to simplify and for comprehensibility, culture is used primarily in the singular in this publication.

Culture does not mean everything. If culture was understood as 'everything', any piece of HCI research would have something to do with it. This assertion is not wrong. One of the seminal definitions of culture argues that culture is 'the most complex whole' (Tylor, 1871). The usefulness of this definition has been challenged though, because it is difficult to operationalize (du Gay et al., 2003). It is very hard (if not impossible) to write a synthesis that brings together HCI and culture when culture means everything. Thus, this book examines HCI research that addresses explicitly culture.

Two meanings of culture, a *particular way of life*, and the *production and circulation of meaning* stand out in the discourses of culture in the literature (Baldwin et al., 2006; Bennett et al., 2005; de Mooij, 2011; du Gay et al., 2003; Hall, 1973; Haviland et al., 2011; Heine, 2016; Hofstede et al., 2010; Schein, 2004; Sun, 2012). We live our lives differently. We have disparate assumptions, customs, religions, taboos, values, and so on. We have dissimilar ways of perceiving the world, relating to and talking to each other, and solving problems. These different lifestyles, activities, behaviors, and ideals are learned and shared among most of the members of a cultural group and are also made by them. The meaning of these particular ways of life is not given to us ready-made; it only becomes available to those who participate in their production and circulation and understand it. Both meanings of culture are particularly relevant for HCI (Sun, 2012), as we aspire to understand the world/s of our users, who are active meaning-makers,

---

[1] The notion of using sensitizing concepts in empirical research was developed by Blumer (cited in Puddephatt et al., 2009, p. 19).

and articulate this knowledge into the design of technologies, which are always socio-culturally constructed artifacts (Bijker, 1995; Harper et al., 2007; Oudshoorn & Pinch, 2003).

This understanding of users and technology is incomplete without considering issues of *identity* and *power*. While culture is often associated with nationality, people integrate multiple cultural identities (Smith-Jackson et al., 2014). We can hear people speak of the Deaf culture (Blume, 2010; Ladd, 2003), the culture of organizations (Sackmann, 2021), the geek culture (McArthur, 2009), or the retrogaming culture in digital gaming (Mora-Cantallops et al., 2021). What makes these groups arguably qualify as 'cultures' is that their members exist within a shared context (e.g., a common history, similar values, and beliefs, a meaningful activity that they do together), communicate with each other, have some norms that create and learn from each other, and that distinguish them from other groups and have some common practices and ideas (Heine, 2016). Culture is always implicated in matters of power. For example, one of the original uses of the term (i.e., intellectual development), which is still predominant today, reinforces the segregation of classes, e.g., the upper versus the working class, and populations, more or less civilized/developed societies. More recently, the widespread applications of algorithms and Artificial Intelligence (AI), which is predominantly conceived and portrayed as White (Cave & Dihal, 2020), have led to significant concerns about how the culture of AI (Elliott, 2019) is amplifying and perpetuating inequity and privilege in society (Umoja, 2018; Obermeyer et al., 2019; Benjamin, 2019). Taking into account cultural identities and power can potentially help HCI to better understand the relationship between digital technologies and a highly diverse user base, and inform the development of less biased and segregating digital technologies.

This view of culture has some overlaps with the notion of context in HCI. As discussed in Dourish (2004), context is not something that describes a setting, such as a location or a time; it is something that people do. This view of context (as an interactional problem) places a strong emphasis on meaning and interaction. Both aspects are shared with culture. What distinguishes culture from context is that culture is a property of people whereas context "is a property of (occasions of) interaction" (Dourish, 2004, p. 26). Culture encompasses, shapes, and goes beyond interaction. It includes visible and non-visible manifestations of ways of living, behaving, feeling, thinking…of being human. People with different cultural backgrounds can participate, create and share the same interactional context, and this context can in turn encourage the participants to reflect on important elements of culture, such as values and assumptions.

Finally, culture and technology are strongly related to each other. As discussed in Hwang et al. (2015), in Western culture(s) there is a tendency to think of digital technologies as tools designed to enable us to work more efficiently and better negotiate our domestic and networked social lives. Such conventions render digital technologies transparent and invisible. (Norman, 1998) argues that technology should be invisible, hidden from sight, so that users can pay more attention to their tasks. This discourse can defer our realization that technology is always with us, everywhere and in everything, and that has an interdependent relationship with us. Technologies are cultural artifacts. Technologies are laden with human, cultural, and social values (Harper et al., 2007). Technologies shape our everyday lives; they are more than tools (e.g., Airoldi, 2022; Norman, 2004; Turkle, 1995, 2011), and as such, sometimes they can (or should) become visible, as they tell us something about ourselves (Marcus et al., 2017; Eukan, 2000; Hwang et al., 2015; Hassenzahl, 2010; Norman, 2004). This book does not view culture (or technology) as separated from each other; rather, it sees them both mutually constitutive of each other.

1.3 Main Contributions

This book argues that culture is a critical dimension in HCI. The study of people minus culture (P–C) does not necessarily produce a more basic understanding of human beings. Instead, it produces an understanding of *something* that lacks recognizable qualities of human existence (Shore, 1996). In designing new technologies, we are not simply providing better tools for getting things done in a previously existing world (Winograd, 1997). We are creating new worlds wherein computers actively contribute to shape them (Airoldi, 2022; Striphas, 2015) and become our digital partners (Grudin, 2017). The importance of culture can also be seen when it stands as an obstacle towards the use and appropriation of technologies (Hodgson et al., 2013; Kyriakoullis & Zaphiris, 2016; Leidner & Kayworth., 2006; Olasina & Mutula, 2015). Common examples in the literature are meaningless icons or metaphors (e.g., a postbox and desktop) for some cultures, colors with different meanings depending on the national culture (Pappachan & Ziefle, 2008; Russo & Boor, 1993), and mismatches between automation technologies and their operators (Hodgson et al., 2013; Leidner & Kayworth., 2006). Digital technologies are often designed under the implicit assumption that members of different cultures use the technology in the same way as designers do, or equally view a given functionality as appropriate for carrying out a given act (Hodgson et al., 2013; Soro et al., 2017; Vatrapu, 2010). However, this is seldom the case. Technology use is a situated (Suchmann, 1987), everyday cultural activity.

This book provides a structured overview of a large body of HCI research on (and with) culture. This overview addresses the main conceptual perspectives of culture within HCI. The Cultural Dimensions of Hofstede (2010), the cross-cultural theory of communication of Hall (1973, 1976, 1982), and the cultural cognitive systems of thought of Nisbett (2003) predominate in much of this HCI research. The overview also addresses the question of how culture has been operationalized in HCI. The book brings together individual analyses of different periods of research and sources to provide an integrated profile of investigation. Much research is quantitative, conducted with university students in few nations, cross-cultural/comparative, and leaves the term culture predominantly without any definition (Clemmensen & Roese, 2010; Kamppuri et al., 2006; Linxen, Cassau, &

Sturm, 2021). The two main approaches to address culture in HCI are also outlined and illustrated with studies adopting them. The *taxonomic perspective* renders human ways of living and thinking into a finite set of elements sorted into categories. The *contingent perspective,* however, looks at culture as a dynamic, constructed, and interactional phenomenon (Halabi & Zimmermann, 2019). This book also argues that culture has largely been used in HCI to (1) understand and account for the ways in which people interact with digital technologies, (2) summarize the ways in which groups of people distinguish themselves from other groups, (3) deal with and reduce the complexity of people's interactions with technologies and (4) find inspiration and foster reflection.

This book puts forward some open issues to encourage further research. HCI research views culture from two (very) different perspectives: culture as a definitive or a sensitizing concept. Most of the studies tend to confirm previous conceptual perspectives of culture. The theoretical works that guide much HCI research on culture were mostly published in the twentieth century. None of them was developed for understanding the relationship between people and digital technologies. Future research could reflect further on these theoretical aspects, which are an important step for research progress. Much HCI research on culture is conducted with students and equates culture with national groups. Future research could consider other user groups and perspectives of culture to embrace further diversity. Virtually all studies have found cultural differences. Yet, what leads to these differences is less clear. Also, what exactly we have in common as far as digital technology use is concerned has received little research attention. By addressing both differences and similarities, HCI research could develop a more profound and complete understanding of the impact of culture on technology design and use, and vice versa. Culture as a stereotyping mechanism and challenges in designing technologies for culturally diverse users are also discussed.

1.4   Overview

The remainder of the book is organized into six chapters. Chapter 2 aims to serve as a backdrop to the book. It traces the origin and development of the term culture. Chapter 2 highlights the importance and complexity of culture, and outlines some of its key traits and ingredients. Chapter 2 presents three main perspectives on the concept of culture across disciplines and also discusses what culture is not.

Chapters 3, 4, and 5 provide an overview of HCI research on culture. This overview aims to address the *why—what—how* questions of culture in HCI research. Chapter 3 addresses the question of *why* culture is important in HCI. Chapter 3 provides a number of reasons for and against the relevance of culture in HCI. Chapter 4 focuses on the *what* question of HCI research on culture. Chapter 4 outlines the main conceptual perspectives of culture within HCI. Chapter 5 examines the ways (i.e., *how*) in which culture has been operationalized in HCI research. It presents a profile of research, the two main research approaches adopted and examples, and the main functions of culture in HCI.

Chapters 6 and 7 wrap up the book. Chapter 6 takes stock of the previous chapters and discusses some open issues. These issues touch upon several aspects, from theory to practice, and are intended to spark discussion and future research. Chapter 7 concludes the book with its key points and some reflections.

The chapters are written so that they can be read (and used) mostly independently. Each chapter has its own references section. However, some chapters draw on previous ones, especially those chapters that present an overview of HCI research on culture (Chaps. 3, 4, and 5). Reading them together is recommended.

References

Airoldi. (2022). *Machine habitus. Toward a sociology of algorithms.* Polity

Almohamed, A., & Vyas, D. (2019). Rebuilding social capital in refugees and asylum seekers. *ACM Transactions on Computer-Human Interaction, 26*(6), 1–30. https://doi.org/10.1145/3364996

Aykin, N. (Ed.). (2005). *Usability and internationalization of information technology.* Lawrence Erlbaum Associates Publishers. https://doi.org/10.1109/memb.2006.1578651.

Baldwin, J. R., Faulkner, S. L., Hecht, M. L., & Lindsley, S. L. (Eds.). (2006). *Redefining culture: Perspectives across the disciplines.* Lawrence Erlbaum Associates Publishers. https://doi.org/10.4324/9781410617002.

Benjamin, R. (2019). *Race after technology. Abolitionist tools for the new Jim code.* Polity.

Bennett, T., Grossberg, L., & Morris, M. (Eds.). (2005). *New keywords. A revised vocabulary of culture and society.* Blackwell Publishing Ltd.

Bijker, W. (1995). *Of bicycles, bakelites, and bulbs.* The MIT Press.

Blume, S. S. (2010). *The artificial ear: Cochlear implants and the culture of deafness.* Rutgers University Press.

Brianza, G., Benjamin, J., Cornelio, P., Maggioni, E., & Obrist, M. (2022). QuintEssence: A probe study to explore the power of smell on emotions, memories, and body image in daily life. *ACM Transactions on Computer-Human Interaction, 29*(6), 1–33. https://doi.org/10.1145/3526950

Cave, S., & Dihal, K. (2020). The whiteness of AI. *Philosophy & Technology, 33*, 685–703. https://doi.org/10.1007/s13347-020-00415-6

Clemmensen, T., & Roese, K. (2010). An overview of a decade of journal publications about culture and Human-Computer Interaction (HCI). *IFIP Advances in Information and Communication Technology, 316*, 98–112. https://doi.org/10.1007/978-3-642-11762-6_9

de Mooij, M. (2011). *Consumer behavior and culture. Consequences for global marketing and advertising.* Sage.

Dix, A., Finlay, J., Abowd, G., & Beale, R. (2004). *Human-computer interaction.* Pearson Education.

Dourish, P. (2004). What we talk about when we talk about context. *Personal and Ubiquitous Computing, 8*(1), 19–30. https://doi.org/10.1007/s00779-003-0253-8

du Gay, P., Hall, S., Janes, L., Mackay, H., & Negus, K. (2003). *Doing cultural studies. The story of the Sony Walkman.* Sage.

Elliott, A. (2019). *The culture of AI: Everyday life and the digital revolution* (1st ed.). Routledge. https://doi.org/10.4324/9781315387185.

English-Lueck, J. (2017). *Cultures@siliconvalley.* Standford University Press.

Epp, F. A., Kantosalo, A., & Mekler, E. D. (2022). Adorned in memes: exploring the adoption of social wearables in Nordic Student Culture. In *Conference on Human Factors in Computing Systems—Proceedings.*

Ekuan, K. (2000). *The aesthetics of the Japanese lunchbox.* The MIT Press.

Gaver, B., Dunne, T., & Pacenti, E. (1999, February). Cultural probes. *ACM Interactions,* 21–29.

Grudin, J. (2017). From tool to partner: The evolution of human-computer interaction. In *Synthesis lectures on human-centered informatics.* Morgan and Claypool. https://doi.org/10.2200/S00745 ED1V01Y201612HCI035.

Halabi, A., & Zimmermann, B. (2019). Waves and forms: Constructing the cultural in design. *AI and Society, 34*(3), 403–417. https://doi.org/10.1007/s00146-017-0713-8

Hall, E. (1973). *The silent language.* Anchor Books.

Hall, E. (1976). *Beyond culture.* Anchor Books.

Hall, E. (1982). *The hidden dimension.* Anchor Books.

Harper, R., Rodden, T., Rogers, Y. & Sellen, A. (Eds.). (2007). Being human: Human computer interaction in the year 2020. https://doi.org/10.1145/1467247.1467265.

Hassenzahl, M. (2010). *Experience design.* Springer.

Haviland, W., Prins, H., McBride, B., & Walrath, D. (2011). *Cultural anthropology.* Wadsworth.

Heine, S. (2016). *Cultural psychology* (3rd ed.). W. W. Norton & Company.

Hertzum, M. (2018). Commentary: Usability—A sensitizing concept. *Human-Computer Interaction, 33*(2), 178–181. https://doi.org/10.1080/07370024.2017.1302800

Hodgson, A., Siemieniuch, C. E., & Hubbard, E.-M. (2013). Culture and the safety of complex automated sociotechnical systems. *IEEE Transactions on Human-Machine Systems, 43*(6), 608–619. https://doi.org/10.1109/THMS.2013.2285048

Hofstede, G., Hofstede, G. J., & Minkov, M. (2010). *Cultures and organizations.* McGraw Hill.

Hwang, I. D., Guglielmetti, M., & Dziekan, V. (2015). Super-natural: Digital life in eastern culture. SIGGRAPH ASIA 2015 Art Papers, 1–7. https://doi.org/10.1145/2835641.2835644.

Kamppuri, M. (2012). Because deep down, we are not the same: Values in cross-cultural design. *Interactions, 19*(2), 65–68. https://doi.org/10.1145/2090150.2090166

Kamppuri, M., Bednarik, R., & Tukiainen, M. (2006). The expanding focus of HCI: Case culture. *ACM International Conference Proceeding Series, 189*(October), 405–408. https://doi.org/10.1145/1182475.1182523

Kyriakoullis, L., & Zaphiris, P. (2016). Culture and HCI: A review of recent cultural studies in HCI and social networks. *Universal Access in the Information Society, 15*(4), 629–642. https://doi.org/10.1007/s10209-015-0445-9

Ladd, P. (2003). *Understanding deaf culture.* Multilingual Matters.

Leidner & Kayworth. (2006). Review: A review of culture in information systems research: Toward a theory of information technology culture conflict. *MIS Quarterly, 30*(2), 357. https://doi.org/10.2307/25148735.

Linxen, S., Cassau, V., & Sturm, C. (2021). Culture and HCI: A still slowly growing field of research. Findings from a systematic, comparative mapping review. In *ACM International Conference Proceeding Series.* https://doi.org/10.1145/3471391.3471421.

Marcus, A., Kurosu, M., Ma, X., & Hashizume, A. (2017). Cuteness engineering. *Springer International Publishing.* https://doi.org/10.1007/978-3-319-61961-3

McArthur, J. A. (2009). Digital subculture: A geek meaning of style. *Journal of Communication Inquiry, 33*(1), 58–70. https://doi.org/10.1177/0196859908325676

Mora-Cantallops, M., Muñoz, E., Santamaría, R., & Sánchez-Alonso, S. (2021). Identifying communities and fan practices in online retrogaming forums. *Entertainment Computing, 38*, 100410. https://doi.org/10.1016/j.entcom.2021.100410.

Nisbett, R. (2003). The geography of thought. *The Free Press*. https://doi.org/10.1111/j.0033-0124. 1973.00331.x

Norman, D. (1998). *The invisible computer*. The MIT Press.

Norman, D. (2004). *Emotional design*. Basic Books.

Obermeyer, Z., Powers, B., Vogeli, C., & Mullainathan, S. (2019). Dissecting racial bias in an algorithm used to manage the heath of populations. *Science, 366*, 447–453.

Olasina, G., & Mutula, S. (2015). The influence of national culture on the performance expectancy of e-parliament adoption. *Behaviour & Information Technology, 34*(5), 492–505. https://doi.org/10.1080/0144929X.2014.1003326

Oudshoorn, N., & Pinch, T. (Eds.). (2003). *How users matter. The co-construction of users and technologies*. The MIT Press. https://doi.org/10.1353/tech.2006.0041.

Pappachan, P., & Ziefle, M. (2008). Cultural influences on the comprehensibility of icons in mobile–computer interaction. *Behaviour & Information Technology, 27*(4), 331–337. https://doi.org/10.1080/01449290802228399

Puddephatt, A., Shaffir, W., & Kleinknecht, S. (Eds.). (2009). *Ethnographies revisited. Constructing theory in the field*. Routledge.

Rodríguez, I., Puig, A., Tellols, D., & Samsó, K. (2020). Evaluating the effect of gamification on the deployment of digital cultural probes for children. *International Journal of Human-Computer Studies, 137*, 102395. https://doi.org/10.1016/j.ijhcs.2020.102395

Russo, P., & Boor, S. (1993). How fluent is your interface? Designing for International Users. *Interchi*, 342–347.

Sabie, D., Ekmekcioglu, C., & Ahmed, S. I. (2022). A decade of international migration research in HCI: Overview, challenges, ethics, impact, and future directions. *ACM Transactions on Computer-Human Interaction, 29*(4), 1–35. https://doi.org/10.1145/3490555

Sackmann, S. A. (2021). *Culture in organizations: Development*. Springer International Publishing. https://doi.org/10.1007/978-3-030-86080-6

Schein, E. (2004). *Organizational culture and leadership*. Wiley.

Sharp, H., Rogers, Y., & Preece, J. (2019). *Interaction design: Beyond human-computer interaction*. Wiley.

Shneiderman, B., Plaisant, C., Jacobs, S. M., & Elmqvist, N. (2018). *Designing the user interface. Strategies for effective human-computer interaction*. Pearson Education.

Shore, B. (1996). *Culture in mind. Cognition, culture, and the problem of meaning*. Oxford University Press.

Singh, N., & Pereira, A. (2005). *The culturally customized web site*. Elsevier.

Smith-Jackson, T. L., Resnick, M. L., & Johnson, K. (Eds.). (2014). *Cultural ergonomics. Theory, methods and applications*. CRC Press.

Soro, A., Brereton, M., Taylor, J. L., Lee Hong, A., & Roe, P. (2017). A cross-cultural notice-board for a remote community: Design, deployment, and evaluation. In R. Bernhaupt, G. Dalvi, A. Joshi, D. K. Balkrishan, J. O'Neill, & M. Winckler (Eds.), *Human-computer interaction—INTERACT 2017* (Vol. 10513, pp. 399–419). Springer International Publishing. https://doi.org/10.1007/978-3-319-67744-6_26.

Striphas, T. (2015). Algorithmic culture. *European Journal of Cultural Studies, 18*(4–5), 395–412. https://doi.org/10.1177/1367549415577392

Sun, H. (2012). *Cross-cultural technology design. Crafting culture-sensitive technology for local users*. Oxford Series in Human-Technology Interaction.

Suchmann, L. (1987) *Plans and situated actions: The problem of human-machine communication*. Cambridge University Press.

Tredinnick, L. (2008). *Digital information culture: The individual and society in the digital age*. Chandos Publishing.

Tukiainen, S. S. I. (2010). Coping with cultural dominance in cross cultural interaction. In *Proceedings of the 3rd ACM International Conference on Intercultural Collaboration, ICIC'10* (pp. 255–258). https://doi.org/10.1145/1841853.1841901.

Turkle, S. (1995). *Life on the screen. Identity in the age of the Internet*. Simon & Schuster.

Turkle, S. (2011). *Alone together*. Basic Books.

Tylor, E. (1871). *Primitive culture*. Cambridge University Press.

Umoja, S (2018). *Algorithms of oppression: How search engines reinforce racism*. New York University Press.

Vatrapu, R. K. (2010). Explaining culture: An outline of a theory of socio-technical interactions. In *Proceedings of the 3rd ACM International Conference on Intercultural Collaboration, ICIC'10* (pp. 111–120). https://doi.org/10.1145/1841853.1841871.

Winograd, T. (1997). From computing machinery to interaction design.

The Concept of Culture: A Short and Guided Overview

2

This chapter presents a short and guided overview of the concept of culture. This chapter offers some background on the origin and development of the term culture. It also outlines some of its key traits and ingredients and summarizes three main perspectives of culture across disciplines. This chapter has two objectives. It aims to serve as a background to discussions on culture in the following chapters of this book. It also aims to serve as a *port of entry* into culture for those HCI scholars who are interested in delving into it. This chapter draws on relevant works in disciplines where culture plays a relevant role, such as Anthropology, Sociology, Cultural Studies, Management, and Cultural Psychology.

2.1 Origin and Development; from Culture to Cultures

As reviewed in (Bennett et al., 2005), culture has its roots in the idea of tending or cultivating crops and animals—as, for example, in agri*culture*. From this original interpretation, we can derive one of its central modern meanings: culture as the process of human development (du Gay et al., 2003). Culture was initially used to describe a general process of intellectual, spiritual, and aesthetic development of individuals. This usage was later on extended, with the expansion of colonization and ethnographic studies, to the growth, progress, and sorting of societies (e.g., less or more civilized). Today, the meaning of culture referring to the high arts, intellectual tradition, and different levels of evolution and progress of societies, persists (Bennett et al., 2005; du Gay et al., 2003). For example, using the term culture to refer to arts prevails among cultural heritage studies (Cardoso et al., 2020; Gou, 2021; Lu et al., 2019; Not & Petrelli, 2019; Petrelli, 2019; Ruiz-Calleja et al., 2023). Yet, social movements and revolutions with their pursuits, which are often different from those of the elite, such as 'the sixties' (Marwick, 1999), growth and diversity of societies (e.g., different lifestyles), and an increasing role

S. Sayago, *Cultures in Human-Computer Interaction*, Synthesis Lectures on Human-Centered Informatics, https://doi.org/10.1007/978-3-031-30243-5_2

of democratic and egalitarian sentiment have given way to another perspective of culture (Bennett, et al., 2005). The term is used to describe *particular ways of life* which express certain *meanings* and values of groups[1] of people, nations, or periods (du Gay et al., 2003).

We talk now about cultures, in the plural. The term culture is used in a wide array of contexts, from uses relating to forms of identity (e.g., Deaf culture (Blume, 2010; Ladd, 2003), youth culture (Hodkinson & Deicke, 2007) and discrimination (Lee & Rich, 2021)), food cultures (Bell, 1997; Gayler et al., 2022), Artificial Intelligence and its role in society (e.g., cultural robotics (Ornelas et al., 2022; Samani et al., 2013), the culture of AI (Elliott, 2019), software development (e.g., agile culture (Beyer, 2010), and the maker culture (Eckhardt et al., 2021; Hatch, 2014), to name a few. We also talk about subcultures, such as retrogaming in digital gaming (Mora-Cantallops et al., 2021) or the geek culture in Internet (McArthur, 2009), which are a subversion to normality or the parent culture that share common interests, and countercultures, i.e., groups of people whose way of life and ideas are opposed to those accepted by mainstream cultures (Bell, 2002; Hebdige, 1991; Hodkinson & Deicke, 2007). Even though the members of a cultural group share an overarching culture, this does not mean that each of them must do and think in the same way, or that they cannot belong to other cultural groups or subgroups. People can have hybrid cultural identities and backgrounds (Burke, 2009). This widespread use of culture shows the importance, complexity, and diversity of the term, as Sect. 2.2 discusses.

2.2   Importance, Complexity, and Diversity

The fact that a concept is shared across different contexts denotes that it brings something significant to discussions of the complex and rich nature of our lives (Baldwin et al., 2006). This in turn implies that any operational definition of culture necessarily focuses on particular aspects and overlooks others. A seminal study carried out by Kroeber and Kluckhohn (1985) presented more than 150 different definitions of culture (Kroeber & Kluckhohn, 1985). An updated collection in 2006 (Baldwin et al., 2006) presents c.300 definitions, which are classified into 7 types of themes, depending on whether culture is seen as (see Table 2.1).

Some definitions of culture—taken from (Baldwin et al., 2006)—are:

- Culture is then properly described not as having its origin in curiosity, but as having its origin in the love of perfection; it is a study of perfection" (p. 34). "Culture is considered not merely as the endeavor to see and learn this, but as the endeavor, also,

---

[1] It might be worth noting that the term group is used rather than, say, crowd to stress the fact that not every collection of people develops a culture (Schein, 2004). A fairly stable membership and a history of shared learning are deemed key to the formation of some degree of culture.

Table 2.1 Seven types of themes of culture (Baldwin et al., 2006)

• *structure*: definitions that look at culture in terms of some pattern or structure of regularities, such as patterns of behavior, ceremonies, rituals, and social organization

• *function*: definitions that focus on what culture does, such as providing people with a shared sense of identity/belonging and means of control over other individuals or groups

• *process*: definitions that focus on the ongoing social construction of culture, such as sense-making, relating to others, and transmitting a way of life

• *product*: *definitions in terms of* objects with symbolic meanings, such as artifacts, buildings, and cultural texts

• *refinement*: definitions that frame culture as intellectual refinement (of individuals or groups)

• *group membership*: definitions that focus on the participation in a collective that shares an understanding of the world

• *power/ideology*: definitions based on group-based power, such as subcultures

---

to make it prevail, the moral, social, and beneficent character of culture be- comes manifest. (Baldwin et al., 2006, p. 143)

- "What really binds men together is their culture—the ideas and the standards they have in common" (p. 16). "A culture, like an individual, is a more or less con- sistent pattern of thought and action" tied to the "emotional and intellectual mainsprings of that society" (Baldwin et al., 2006, p. 147).

- Organizational culture is the norms, attitudes, values, beliefs, and philosophies of an enterprise. (Baldwin et al., 2006, 171)

- Cultures are highly specific systems that both explain things and constrain how things can be known. (Baldwin et al., 2006, p. 202)

This diversity is seen from two opposing viewpoints in the literature (Bennett et al., 2005). One embraces variety, assumes that definitions of culture are contextual and dis- cursive (Baldwin et al., 2006) and that this fragmentation and contestation is part of the scientific work (Kuhn, 1996). This point of view regards culture as a sensitizing concept (Blumer, cited in Puddephatt et al., 2009, p. 19). The other hesitates over the value of the term culture, because it is a highly compromised idea (Geertz, 1973), without a unified and common definition. This perspective sees culture as a definitive (i.e., well-defined) concept. These two (very different) perspectives are discussed further in Chap. 6.

Notwithstanding this diversity, there is widespread agreement that culture is learned, constructed, and shared. Culture gives order and meaning to our lives as well as the lives of others, and it is omnipresent. Even without our being consciously aware of it, culture strongly influences how we think, communicate, and behave. The following Sects. 2.3–2.5 address these key ingredients of culture.

2.3   Key Ingredients of Culture

2.3.1   Culture is Learned, Shared, and Emerges in Interaction

For something (any kind of idea, belief, technology, habit, or practice) to be considered culture, it must be learned as well as shared. Understanding what is cultural involves two parts (Haviland et al., 2011). One of them consists of separating what is shared from what is very individually variable. Culture represents what is shared (i.e., a common denominator) within a group, and presumably not shared (or not entirely shared) outside it (Wallerstein, 1990). The 'cultural shock' we experience when we work and live in another culture is an example of this common denominator. Moving to a new or different culture involves adjustment (language, social behaviors, etc.). We also need to understand whether common behaviors and ideas are learned. Culture is transmitted across periods and generations. We acquire and absorb culture through learning from others (Heine, 2016).

Another key characteristic of culture is that it emerges in *interaction*. We are constantly reinforcing culture, or building new elements of it (such as artifacts, values, practices, beliefs, and whatever else), as we encounter new people and experiences (Schein, 2004). Culture emerges in adaptive interactions between humans and environments (Triandis, 2007). Culture is enacted and created by our interactions with others.

2.3.2   Humans Cannot Do Without Culture

We make numerous decisions throughout our lives. Yet, we do not have the option to be *acultural* (Rose & Yam, 2007). We do not have the option to be asocial either. That is, living as a neutral individual, who is not bound to particular practices and socio-culturally structured ways of behaving, is not available.

Interdisciplinary research shows that culture influences the way we eat (Bell, 1997), sleep (Heine, 2016), relate to each other (e.g., individualism or collectivism) (Triandis, 1995), perceive the world and think (Hall, 1973, 1976, 1982; Nisbett, 2003; Nisbett & Norenzayan, 2001), behave (Kitayama & Cohen, 2007; Keith, 2019, learn (Rogoff, 2003; Gay, 2018), and develop (Heine, 2016; Keith, 2019; Kitayama & Cohen, 2007; Rogoff, 2003). Some examples of this influence of culture are summarized next.

People who grow up and live in 'carpentered' environments, i.e., squared, city-block environments and rectangular buildings are more susceptive to the Müller-Lyer illusion (Phillips, 2019, p. 289). Two horizontal lines of exactly the same length, one with arrow heads pointing outwards, and the other with arrowheads pointing inwards, appear to be different in length.[2]

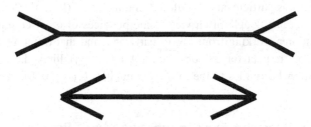

What is funny in some cultures might not be seen as that funny in other cultures (Heine, 2016). This is something the reader is likely to have experienced if s/he has lived abroad, watched foreign comedies, or have friends and colleagues from other countries.

Some societies (eg., Mexican) hold ceremonies that acknowledge the onset of puberty, i.e., the beginning of adolescence, while in others puberty is rarely discussed or celebrated (Gibbons & Poelker, 2019). Cultural taboos and limiting social norms in some countries like India make it challenging to communicate and teach about menstrual health whereas in other societies these cultural taboos do not exist or are not so strong (Tuli et al., 2019).

The heart of who we think we are varies in important ways across cultures. People from individualist cultures tend to be more self-contained and individual (e.g., they view themselves as distinct from all others) than those from collectivist cultures, who focus more on connection with others (e.g., roles) and view in-group members as an extension of themselves (Heine, 2016).

Depending on your culture (national group), what is different is dangerous, rules are important, even if the rules will never work, and there is only one truth. On the contrary, it might well be that what is different causes curiosity, rules should be limited to those that are necessary, and deviant and innovative ideas are tolerated (Hofstede, 2002; Hofstede et al., 2010).

Westerners tend to perceive the world analytically, seeing objects as discrete and separate, whereas East Asians are more likely to perceive the world in holistic terms, stressing the relations between objects (e.g., Nisbett & Norenzayan, 2001). These differences are mostly due to the socio-cultural environments in which these groups of people are born and raised (Nisbett, 2003).

The question of whether emotional experiences are similar or different across cultures is a challenging one (Heine, 2016; Lottridge et al., 2012). Strong arguments have been made for both cases, i.e., emotions are experienced identically around the world

---

[2] The segment with arrow heads pointing outwards appears longer than the segment with arrow heads pointing inwards—see https://michaelbach.de/ot/sze-muelue/index.html (Accessed 11 January, 2023).

(Ekman, 2003) or that cultural experiences determine the kinds of emotions one has (Barrett, 2017).[3]

Humans are born both as members of their species and as members of their communities, and develop as participants in cultural communities (Rogoff, 2003). This might account for the fact that people often view some practices of other communities as barbaric, such as children working from a very early age, and that different people learn in different ways, e.g., respect for authority, orality, and storytelling, the relevance of the content of educational activities to the local communities (Gay, 2018).

2.3.3   Culture Does Not Exist Independently of People

Culture might be seen as a 'thing' out there because it surrounds us at all times. We live our lives differently, speak different languages, have different customs, eat different foods, and have different religious beliefs and values. We are not aware of most of these differences until we experience a different culture from ours. For us, thinking and behaving in the way we do is just 'normal'. These different lifestyles, activities, behaviors, and ideals are learned and shared among most of the members of a cultural group. Yet, culture is not given to us ready-made (Shore, 1996). It is us (people) who create culture. Thus, to study culture we need to consider the people who actively construct their worlds and give meaning to them (Kitayama & Cohen, 2007).

2.4   Cultural Change and Persistence

Cultures are not static entities. When our parents (and grandparents) were younger, they likely participated in a quite distinct (national) culture from what we do today. Why cultures vary defies any single answer (Heine, 2016). Culture is learned and shared, so we are 'prisoners' of the culture we inherit. Yet, the decisions we make as a group, and the outcomes of those decisions, lead to cultural evolution. Changes in a culture often happen because people realize that certain old ways of doing things or thinking do not work anymore (Sackmann, 2021; Schein, 2004; Trompenaars & Hampden-Turner, 1997). Changes also might happen in response to events like population, aging, and environmental and

---

[3] A recent large-scale study (Srinivasan & Martinez, 2021) of the production and visual perception of facial expressions of emotion in the wild (7.2 million images, 10 K hours of video) found that of the 16 K possible facial configurations that people can produce only 35 are successfully used to transmit affect information across cultures, and only 8 within a smaller number of cultures. This study also found that the degree of successful visual interpretation of these 43 expressions varies significantly, and the number of expressions used to communicate each emotion is also different, i.e., 17 expressions transmit happiness, but only 1 is used to convey disgust. This means that computer vision algorithms may be useful in some applications (searching for pictures in a digital photo album) but not in others (in the courts).

health crises (Haviland et al., 2011; Katz, 2009). These events encourage us to re-think assumptions about, for instance, aging (e.g., a period of losses, gains, or a combination?), mobility (e.g., less contamination, car sharing, public transport), and working styles, (e.g., working from home and videoconferencing since the COVID-19 pandemic). Changes also occur because of the introduction and use of digital technologies in the everyday lives of most of us, like social media and algorithms, which shape connectivity (Van Dijck, 2013) and hold society together (Airoldi, 2022).

In the face of change, how do cultures persist? The thread of every cultural innovation must be woven into something, and this something is an existing (and inherited) web of beliefs and practices (Heine, 2016). To simplify, let me consider culture as a national group. There are reasons to think that 22nd-century Spanish culture will (in many ways) be different from 21st-century Spanish culture. If we look back, there have been many— and some of them, profound – changes since the Spanish civil war (e.g., democracy, the Catalan is not forbidden, women play many different roles in society other than taking care of their families, and so on). Culture changes, because people make it change. Yet, 22nd-century Spanish will remain distinctively Spanish.

2.5 Overarching Perspectives of Culture

The previous sections of this chapter have outlined relevant aspects of the concept of culture. This section summarizes three main perspectives of the concept of culture across disciplines (Baldwin et al., 2006).

2.5.1 Culture as Meaning—The Interpretive Perspective

One of the most prominent views of culture, especially in Anthropology, around which the whole discipline arose, sees culture as meaning. The work of Clifford Geertz is highly influential in this vision of culture, "(I) take(s) culture to be webs of significance, and the analysis of it to be not an experimental science in search of law but an interpretive one in search of meaning" (Geertz, 1973). The research question to ask within this view is what the import of the object of study (e.g., an artifact, a practice) is rather than its ontological status (i.e., what it is). By doing so, this vision assumes that people are active meaning-makers and the main task of the researchers is to interpret, i.e., understand and explain, the particular ways (patterns) in which the meaning of something is co-created and shared (Geertz, 1973). Ethnography plays a key role in gaining this understanding (Fetterman, 2010).

In order to follow a baseball game, one must understand what a bat, a hit, an inning, a left fielder, a squeeze play, a hanging curve, and a tightened infield are, and what the game in which these "things" are elements is all about (Geertz, 1983).

2.5.2   Culture as Identity—The Intergroup Perspective

Whereas the interpretive perspective casts culture as meaning, the intergroup view focuses on membership and identity. This vision shifts the focus on culture from fixed categories such as geography, biology, and nation–states to a definition based on group membership. To this end, one issue of importance in the study of culture is how to specify its limits (Triandis, 2007). One way of doing so is to treat culture as synonymous with a demarcated population defined with respect to certain demographic, geographic, national, or ethnic characteristics. While this view has been (and continues to be) instrumental in studying culture, probably because it is easy to operate, there are other kinds of groups aside from national and regional boundaries that can be argued to have cultures. We can hear people speak of the Deaf culture (Blume, 2010; Ladd, 2003), the maker culture (Eckhardt et al., 2021; Hatch, 2014), the Silicon Valley cultures (English-Lueck, 2017), and the Nordic student culture (Epp, Kantosalo & Mekler, 2022). What makes these groups arguably qualify as 'cultures' is that their members exist within a shared context, communicate with each other, have some norms that create and learn from each other, and that distinguish them from other groups, and have some common practices and ideas (Heine, 2016).

2.5.3   Culture as Power—The Critical Perspective

The Cultural Studies approach defines culture as the creation of hierarchy through ideology. Within this vision of culture, the key meanings are about power and dogma. Cultural studies ask questions about which meanings are put into circulation, by whom, for what purposes, and in whose interests (Barker & Jane, 2016, p. 84) Culture is seen to exist as a means of one group exerting dominance (political, social, artistic, ideational) over others (Wallerstein, 1990). Culture is always implicated in matters of power because power is present in all social arrangements. The Disability culture is a noteworthy example:

We share a common history of oppression and a common bond of resilience. We generate art, music, literature, and other expressions of our lives and our culture, infused from our experience of disability. Most importantly, we are proud of ourselves as people with disabilities. We claim our disabilities with pride as part of our identity. We are who we are: we are people with disabilities (…) It is absolutely not our job to fit into mainstream society. Mainstream society needs to figure not how we fit in, but how we can be of benefit exactly the way we are. (Editorial, 2001)

2.6   **What Culture Is Not**

Culture is not synonymous with nationality. Although nationality and culture are commonly overused, which implies national uniformity, we need to look only at such culturally diverse nations as Spain to see that a nation may include many cultural and subcultural groups (the Catalan, Andalusians, or Basques are just some examples). People might also belong simultaneously to multiple cultures (e.g., the culture of the organization they work for (Sackmann, 2021)) regardless of their national nationality (English-Lueck, 2017; Epp et al., 2022; Tukiainen, 2010). Yet, within this diversity, people who share the same language, traditions, characteristics, religion, history, etc. can display a large amount of stability in, for instance, their values (e.g., the group is more important than the individual or vice-versa), beliefs (e.g., 'he who proposes, does'), or how they perceive the world (e.g., holistic or analytic) (Hofstede et al., 2010; Nisbett, 2003). The problem lies in endorsing (or assuming) national cultural determinism (McSweeney, 2002).

Culture is not (exclusively) what you get when you study Shakespeare, listen to classical music, take courses in art history, or visit a museum (Groh, 2020). Laypeople usually associate culture with the 'high arts'. Yet, culture is a vessel (or a fleet of vessels) that people load with meaning from their own academic backgrounds and personal experiences (Baldwin et al., 2006). For example, Cultural Psychology focuses on the extent to which culture shapes how people's minds operate (Heine, 2016). Culture in the context of organizations (Organizational Culture) refers to visible manifestations and basic beliefs commonly held by the members of a group regarding solving problems concerning internal integration and external adaptation that influences their perceptions, thinking, actions and feelings (Sackmann, 2021; Schein, 2004). None of these meanings (or any of the others that can be found in the literature (Baldwin et al., 2006) is arguably wrong or the most correct one. Culture, as a signifier/idea, is a moving target.

Culture is not just what you see. Perhaps the most intriguing aspect of culture as a concept is that it points us to phenomena that are below the surface (Kamppuri, 2012; Schein, 2004). The iceberg model or metaphor is often used in the literature to illustrate this aspect (see Fig. 2.1). At the top of the iceberg, we see the visible manifestations of cultures, such as observable artifacts (e.g., buildings, furniture) and behavior, both verbal (e.g., language, humor, legends) and nonverbal (e.g., gestures and celebrations). The most significant part of the iceberg lies underneath the surface of water and is invisible. Likewise, the essential components of culture are also hidden, such as norms or standards of good and bad behavior, values (e.g., the value statement of organizations), and assumptions, which are often taken-for-granted and non-debatable, such as criteria for determining truth (e.g., dogma, based on pragmatism, tradition or trust in the authority of leaders). To decipher culture, we both identify and understand the visible and non-visible manifestations of it. For example, the similar scores of Finland and Tanzania in uncertainty avoidance (Hofstede et al., 2010) suggest that members of these countries would feel equally uncomfortable in unknown situations. But in practice, Finnish and

Fig. 2.1 Iceberg model

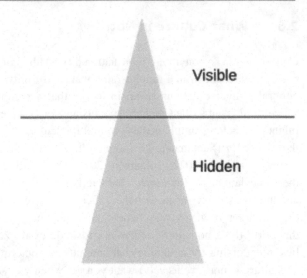

Tanzanian people use different techniques for dealing with the unknown. Finnish observe the situation silently from a safe distance while the Tanzanian strategy is perhaps geared more toward initiating contact and using talk to gather information to reduce uncertainty (Kamppuri, 2012).

Culture is not a collection of artifacts, values, and assumptions disconnected from each other. Why do we need culture if we have artifacts, values, and assumptions? When we say that something is cultural, we imply that it is not only shared, but also learned and stable, because it is a way of summarizing the ways in which groups distinguish themselves from other groups (Wallerstein, 1990). Culture implies that artifacts, values, and assumptions tie together into a coherent whole (Schein, 2004).

References

Airoldi. (2022). *Machine habitus. Toward a sociology of algorithms.* Polity

Baldwin, J. R., Faulkner, S. L., Hecht, M. L., & Lindsley, S. L. (Eds.). (2006). *Redefining culture: Perspectives across the disciplines.* Lawrence Erlbaum Associates Publishers. https://doi.org/10.4324/9781410617002

Barker, C., & Jane, E. (2016). *Cultural studies. Theory and practice.* Sage

Barrett, L. (2017). *How emotions are made. The secret life of the brain.* Houghton Mifflin Harcourt

Bell, D. (1997). *Consuming geographies: We are where we eat.* Routledge.

Bell, D. (2002). *An introduction to cybercultures.* Routledge.

Bennett, T., Grossberg, L., & Morris, M. (Eds.). (2005).*New Keywords. A Revised vocabulary of culture and society.* Blackwell Publishing Ltd.

Beyer, H. (2010). *User-centered agile methods.* Morgan & Claypool Publishers.

Blume, S. S. (2010). *The artificial ear: Cochlear implants and the culture of deafness.* Rutgers University Press

Burke, P. (2009). *Cultural hybridity*. Polity Press.

Cardoso, P. J. S., Rodrigues, J. M. F., Pereira, J., Nogin, S., Lessa, J., Ramos, C. M. Q., Bajireanu, R., Gomes, M., & Bica, P. (2020). Cultural heritage visits supported on visitors' preferences and mobile devices. *Universal Access in the Information Society, 19*(3), 499–513. https://doi.org/10.1007/s10209-019-00657-y

du Gay, P., Hall, S., Janes, L., Mackay, H., & Negus, K. (2003). *Doing cultural studies. The story of the Sony Walkman*. Sage

Eckhardt, J., Kaletka, C., Pelka, B., Unterfrauner, E., Voigt, C., & Zirngiebl, M. (2021). Gender in the making: An empirical approach to understand gender relations in the maker movement. *International Journal of Human-Computer Studies, 145*, 102548. https://doi.org/10.1016/j.ijhcs.2020.102548

Editorial. (2001). *Disability culture—Independent living institute*. Newsletter

Ekman, P. (2003). *Emotions revealed*. Times Books.

Elliott, A. (2019). *The culture of AI: Everyday life and the digital revolution* (1st ed.). Routledge. https://doi.org/10.4324/9781315387185

English-Lueck, J. (2017). *Cultures@siliconvalley*. Standford University Press

Epp, F. A., Kantosalo, A., & Mekler, E. D. (2022). Adorned in memes: Exploring the adoption of social wearables in Nordic Student Culture. In *Conference on Human Factors in Computing Systems—Proceedings*

Fetterman, D. (2010). *Ethnography. Step-by-Step*. Sage

Gay, G. (2018). *Culturally responsive teaching. Theory, research, and practice*. Teachers College Press

Gayler, T., Sas, C., & Kalnikaitė, V. (2022). Exploring the design space for human-food-technology interaction: An approach from the lens of eating experiences. *ACM Transactions on Computer-Human Interaction, 29*(2), 1–52. https://doi.org/10.1145/3484439

Geertz, C. (1973). *The interpretation of cultures*. Basic Books.

Geertz, C. (1983). *Local knowledge. Further essays in interpretive anthropology*. Basic Books. https://doi.org/10.1177/1350507602334002

Gibbons, J., & Poelker, K. (2019). Adolescent development in a cross-cultural perspective. In Keith, K. (Ed.), *Cross-cultural psychology. Contemporary themes and perspectives* (pp. 190–216). Wiley Blackwell

Gou, Y. (2021). Computer digital technology in the design of intangible cultural heritage protection platform. In *2021 3rd International Conference on Artificial Intelligence and Advanced Manufacture* (pp. 1524–1528). https://doi.org/10.1145/3495018.3495434

Groh, A. (2020). *Theories of culture*. Routledge, Taylor & Francis group

Hall, E. (1973). *The silent language*. Anchor Books

Hall, E. (1976). *Beyond culture*. Anchor Books

Hall, E. (1982). *The hidden dimension*. Anchor Books

Hatch, M. (2014). *The maker movement manifesto: Rules for innovation in the new world of crafters, hackers, and tinkerers*. McGrawHill

Haviland, W., Prins, H., McBride, B., & Walrath, D. (2011). *Cultural anthropology*. Wadsworth.

Hebdige, D. (1991). *Subculture: The meaning of style*. Routledge

Heine, S. (2016). *Cultural psychology* (3rd ed.). W. W. Norton & Company

Hodkinson, P., & Deicke, W. (Eds.). (2007). *Youth cultures: Scenes, subcultures and tribes*. Routledge

Hofstede, G. (2002). *Exploring culture. Exercises, stories, and synthetic cultures*. Intercultural Press

Hofstede, G., Hofstede, G. J., & Minkov, M. (2010). *Cultures and organizations*. McGraw Hill.

Kamppuri, M. (2012). Because deep down, we are not the same: Values in cross-cultural design. *Interactions, 19*(2), 65–68. https://doi.org/10.1145/2090150.2090166

Katz, S. (2009). *Cultural aging. Life course, lifestyles, and senior worlds*. University of Toronto Press

Keith, K. (Ed.). (2019). *Cross-cultural psychology. Contemporary themes and perspectives*. Wiley Blackwell

Kitayama, S., & Cohen, D. (Eds.). (2007). *Handbook of cultural psychology*. The Guildford Press

Kroeber, A., & Kluckhohn, C. (1985). *Culture: A critical review of concepts and definitions*. Vintage Books

Kuhn, T. (1996). *The structure of scientific revolutions*. The University of Chicago Press

Ladd, P. (2003). *Understanding deaf culture*. Multilingual Matters.

Lee, M. K., & Rich, K. (2021). Who is included in human perceptions of AI?: Trust and perceived fairness around healthcare AI and cultural mistrust. In *Proceedings of the 2021 CHI Conference on Human Factors in Computing Systems* (pp. 1–14). https://doi.org/10.1145/3411764.3445570

Lottridge, D., Chignell, M., & Yasumura, M. (2012). Identifying emotion through implicit and explicit measures: Cultural differences, cognitive load, and immersion. *IEEE Transactions on Affective Computing, 3*(2), 199–210. https://doi.org/10.1109/T-AFFC.2011.36

Lu, Z., Annett, M., Fan, M., & Wigdor, D. (2019). "I feel it is my responsibility to stream" Streaming and engaging with intangible cultural heritage through livestreaming. In *Conference on Human Factors in Computing Systems—Proceedings* (pp. 1–14). https://doi.org/10.1145/3290605.330 0459

Marwick, A. (1999). *The sixties: Social and cultural transformation in Britain, France, Italy and the United States, 1958–74*. Bloomsbury Reader

McArthur, J. A. (2009). Digital subculture: A geek meaning of style. *Journal of Communication Inquiry, 33*(1), 58–70. https://doi.org/10.1177/0196859908325676

McSweeney, B. (2002). Hofstede's model of national cultural differences and their consequences: A triumph of faith—a failure of analysis. *Human Relations, 55*(1), 89–118. https://doi.org/10.1177/0018726702551004

Mora-Cantallops, M., Muñoz, E., Santamaría, R., & Sánchez-Alonso, S. (2021). Identifying communities and fan practices in online retrogaming forums. *Entertainment Computing, 38*, 100410. https://doi.org/10.1016/j.entcom.2021.100410

Nisbett, R. (2003). The geography of thought. *The Free Press*. https://doi.org/10.1111/j.0033-0124. 1973.00331.x

Nisbett, R. E., & Norenzayan, A. (2001). Culture and systems of thought: Holistic versus analytic cognition. *Psychological Review, 108*(2), 291–310.

Not, E., & Petrelli, D. (2019). Empowering cultural heritage professionals with tools for authoring and deploying personalised visitor experiences. *User Modeling and User-Adapted Interaction, 29*(1), 67–120. https://doi.org/10.1007/s11257-019-09224-9

Ornelas, M. L., Smith, G. B., & Mansouri, M. (2022). Redefining culture in cultural robotics. *AI & SOCIETY*. https://doi.org/10.1007/s00146-022-01476-1

Petrelli, D. (2019). Tangible interaction meets material culture: Reflections on the meSch project. *Interactions, 26*(5), 34–39. https://doi.org/10.1145/3349268

Phillips, W. (2019) Cross-cultural differences in visual perception of color, illusions, depth, and pictures. In Keith, K. (Ed.). *Cross-cultural psychology. Contemporary themes and perspectives* (pp. 287–309). Wiley Blackwell

Puddephatt, A., Shaffir, W., & Kleinknecht, S. (Eds.). (2009). *Ethnographies revisited. Constructing theory in the field*. Routledge

Rogoff, B. (2003). The cultural nature of human development. *Oxford University Press*. https://doi.org/10.1525/jlin.2005.15.2.290

Rose, H., & Yam, M. (2007). Sociocultural psychology: The dynamic interpendence among self systems and social systems. In S. Kitayama & D. Cohen (Eds.), *Handbook of cultural psychology* (pp. 3–40). The Guildford Press.

Ruiz-Calleja, A., Bote-Lorenzo, M. L., Asensio-Pérez, J. I., Villagrá-Sobrino, S. L., Alonso-Prieto, V., Gómez-Sánchez, E., García-Zarza, P., Serrano-Iglesias, S., & Vega-Gorgojo, G. (2023). Orchestrating ubiquitous learning situations about Cultural Heritage with Casual Learn mobile application. *International Journal of Human-Computer Studies, 170*, 102959. https://doi.org/10.1016/j.ijhcs.2022.102959

Sackmann, S. A. (2021). *Culture in organizations: development*. Springer International Publishing. https://doi.org/10.1007/978-3-030-86080-6

Samani, H., Saadatian, E., Pang, N., Polydorou, D., Fernando, O. N. N., Nakatsu, R., & Koh, J. T. K. V. (2013). Cultural robotics: The culture of robotics and robotics in culture. *International Journal of Advanced Robotic Systems, 10*(12), 400. https://doi.org/10.5772/57260

Schein, E. (2004). *Organizational Culture and leadership*. John Wiley & Sons.

Shore, B. (1996). *Culture in mind. Cognition, culture, and the problem of meaning*. Oxford University Press.

Srinivasan, R., & Martinez, A. M. (2021). Cross-cultural and cultural-specific production and perception of facial expressions of emotion in the wild. *IEEE Transactions on Affective Computing, 12*(3), 707–721. https://doi.org/10.1109/TAFFC.2018.2887267

Triandis, H. (1995). *Individualism & Collectivism*. Routledge.

Triandis, H. (2007). Culture and psychology: A history of the study of their relationship. In S. Kitayama & D. Cohen (Eds.), *Handbook of cultural psychology* (pp. 59–77). The Guildford Press.

Trompenaars, F., & Hampden-Turner, C. (1997). *Riding the waves of culture. Understanding cultural diversity in business*. Nicholas Brealey Publishing.

Tukiainen, S. S. I. (2010). Coping with cultural dominance in cross cultural interaction. In *Proceedings of the 3rd ACM International Conference on Intercultural Collaboration, ICIC'10* (pp. 255–258). https://doi.org/10.1145/1841853.1841901

Tuli, A., Dalvi, S., Kumar, N., & Singh, P. (2019). "It's a girl thing": Examining challenges and opportunities around menstrual health education in India. *ACM Transactions on Computer-Human Interaction, 26*(5), 1–24. https://doi.org/10.1145/3325282

Van Dijck, J. (2013). *The culture of connectivity*. Oxford University Press.

Wallerstein, I. (1990). Culture as the ideological battleground of the modern world-system. *Theory, Culture & Society, 7*(2–3), 31–55. https://doi.org/10.1177/026327690007002003

Culture Matters in HCI

3

Why is culture relevant in HCI? This chapter addresses this question by presenting some arguments for and against the importance of culture in HCI. Chapters 4 and 5 also address this question by presenting the main conceptual perspectives of culture within HCI and how this concept has been operationalized in the field. Arguments against include a lack of a universal or common definition of the term culture and globalization. Arguments for touch upon important aspects of HCI, including a diversely growing user base, the need for providing designers with enough support to design across cultures, and the inseparable relationship between culture and technology. This chapter does not aim to discuss all possible reasons for arguing that culture is (not) important in HCI. Rather, it aims to provide research-based arguments for engaging in a discussion. The stance adopted in this chapter and book is that culture matters in HCI.

3.1 Some Arguments Against and for the Relevance of Culture in HCI

There are reasons to be skeptical about the relevance of culture in HCI. For example, Chapter 1 points out that the HCI community has not developed a massive interest in culture. Much research is quantitative, conducted with university students in few nations, cross-cultural/comparative, and leaves the term culture predominantly without any definition (Clemmensen & Roese, 2010; Kamppuri et al., 2006; Linxen, Cassau & Sturm, 2021). Chapter 1 also shows that culture has been partially addressed in some leading works of HCI. Moreover, as opposed to other key dimensions of HCI, such as usability and user experience, a synthesis of culture is largely missing in the HCI literature.[1] Thus,

---

[1] While other books, synthesis, and edited collections have addressed topics so important like usability and user experience (see, for example, this Synthesis Lectures on Human-Centered Informatics),

S. Sayago, *Cultures in Human-Computer Interaction*, Synthesis Lectures on Human-Centered Informatics, https://doi.org/10.1007/978-3-031-30243-5_3

if culture was a main topic of interest in the field of HCI, much more research attention could (or should) have been attracted. Another argument against the importance of culture in HCI is that the term culture defies a single and unique definition (see Chap. 2). In addition to this, culture is constantly changing. These traits of the term culture might encourage us to hesitate over its value in HCI. What does culture bring to the study of HCI if it is so difficult to define and dynamic? Moreover, does it make sense to talk about culture when the world is becoming increasingly 'flat' (Friedman, 2005), i.e., borderless? What is the value of culturally-sensitive technologies in the age of the *global village*? (You et al., 2016).

In light of the above arguments, culture, as a research topic, does not *seem* to be (very) important in HCI. Yet, there are other reasons for arguing just the opposite:

- Culture might not have received the particular attention (some of us think) it deserves as a topic of interest in the HCI community Yet, it does not mean that culture plays a minor role in this field. For example, culture has shaped the way in which user interfaces are designed. Also, Chap. 5 shows that culture is both used as a unit of analysis in a great deal of HCI research and regarded as the main (or one of the mains) determinants of interaction.
- A fundamental tenet of (good) HCI is to 'know the user'. In light of a diversely growing user base, who is the user? Is there a 'global' user? As stated in Chap. 1, culture is a key dimension for understanding us (people). Thus, a cultural understanding of users is needed if digital technologies are to be usable, useful, and appealing to such a wide range of users.
- The concept of usability, which is fundamental in HCI, is widely assumed to be understood similarly across cultures. Yet, usability does not mean the same to all of us (Hertzum, 2010; Winschiers & Fendler, 2007). Also, usability evaluation methods are not conducted in the same way across cultures. These differences are not subjective. They can be accounted for cultural reasons.
- HCI is not only concerned with understanding and evaluating people's interactions with digital technologies but designing them. There is widespread agreement that ensuring those who will use technologies play a critical role in their design is important to design better technologies. Yet, how to deal with culture in Participatory Design is being perceived as an increasingly pressing issue in this domain (Hakken & Maté, 2014).
- Technology and culture are two sides of the same coin (du Gay et al., 2003), as technology shapes culture (e.g., social media technologies and our experience of sociality (Van

---

the author is not aware of any that provides a synthesis of culture in HCI research. Publications that deal with approaches to address culture in HCI, and that examine research on culture across different periods and sources (journals/conference proceedings) have been conducted and are included in this synthesis. These individual works address particular aspects. This synthesis brings them together.

Dijck, 2013)), and culture shapes technology (e.g., the culture of youth and technologies for 'successful' aging (Cordella & Poiani, 2021)). This endless loop is important for HCI to develop a richer understanding of the different ways in which technologies are designed and used, and the roles and meanings of technology in our different ways of living.

These arguments are developed further in the sections that follow.

3.2   Culture Matters in HCI: Some Reasons

3.2.1   Shaping User Interface Design

In the early years of computers (1950–1980s), culture was not a significant issue in the interaction with them. The user base consisted, by and large, of domestic and highly technical users (Marcus, 2001), mostly operators and programmers. There were other more relevant challenges to address than cultural diversity, such as reducing operator burden and constructing better user interfaces (Grudin, 2012). In the 1990s, with the popularization of Graphical User Interfaces (GUIs) and the Internet, there was an increasing technological sophistication in many countries, since computer software products became (almost) a worldwide commodity (Fernandes, 1994). User interfaces went global.

This growing technological sophistication resulted in large world trade (Nielsen, Del Galdo, & Sprung, 1990). Computer software products developed in one country were sold and used by people (with different profiles) in foreign countries. These th ation), as usability for this ree key factors (technological sophistication, global software markets, and diverse user base) encouraged the HCI community to begin to pay attention to culture (in the sense of culture = nation), as usability for this large and diverse number of users depended on increased cultural awareness of issues related to international user interfaces (Del Galdo & Nielsen, 1996; Fernandes, 1994; Nielsen, Del Galdo & Sprung, 1990; Russo & Boor, 1993).

"International user interfaces are those which are intended for use in more than one country. Designing international user interfaces may or may not involve translation, but it should certainly involve consideration of the special needs of other countries and cultures" (Russo & Boor, 1993, p. 291). However, the problem was that little thought had been given to how these user interfaces should be designed (i.e., translated) for use in other nations (Fernandes, 1994). GUIs are a noteworthy example. GUIs adopt metaphors, such as mailboxes, garbage cans, and folders, that are familiar and real world based…but which real world? North American. The point of view taken by most UI researchers and designers was to assume a homogeneous set of end-users with the same cultural basis as them (Nielsen, Del Galdo & Sprung, 1990). As the user base spread, this assumption was challenged. A large number of user interface elements, from symbols, units of measure, and text to time and date formats, and meaning of colors, vary from culture to culture (Del Galdo & Nielsen, 1996; Fernandes, 1994; Nielsen, Del Galdo & Sprung, 1990; Russo & Boor, 1993).

Special consideration was needed for designing international user interfaces. Yet, it was far from easy. The way of doing so involved two main steps: internationalization (i.e., isolating the culturally specific elements from a product) and localization (i.e., infusing a specific cultural context into an internationalized product) (Russo & Boor, 1993). A substantial obstacle to arriving at any feasible solution is simply that we are working with more unknowns than knowns (Nielsen, Del Galdo & Sprung, 1990, p. 294). Translating a user interface is not the same as translating a book (Nielsen, Del Galdo & Sprung, 1990). A user interface should reflect, among other aspects, the values, ethics, and language of the nation in question (Russo & Boor, 1993). Conducting a usability test in a foreign country was a challenge too. In addition to the need of a translator, and the cultural issue (e.g. how to welcome and read the user), the costs associated with running a test were a barrier (Del Galdo & Nielsen, 1996).

Overall, the cultural question was a problem (and it continues to be, as discussed in Chaps. 6 and 7). The *Cultural Dimensions* of Hofstede et al. (2010) introduced in HCI by Marcus and Gould (Marcus & Gould, 2000) in the 2000s was a turning point. The Cultural Dimensions could be applied to the design of global (mostly web at that time) user interfaces and provided user interface designers and researchers with (most of) the knowledge they needed. Since then, culture has played an important role in user interface design, shaping it. As this chapter shows later on, a growing number of efforts are being made to operationalize culture in design, and technology and culture are two sides of the same coin.

3.2.2 A Diversely Growing and Active User Base

As argued by Marcus and Gould (2000), user interfaces for desktop, Web, mobile, and vehicle platforms, to name a few, reach across culturally diverse user communities, sometimes within a single country/language group, and certainly across the globe. At the outset of the twenty-first century, a cultural understanding of users (Alexander et al., 2021) is needed if digital technologies are to be usable, useful, and appealing to such a wide range of users. For example, culture mediates the expectations we all have for people with social and intellectual disabilities and digital technologies (Mankoff, Hayes, & Kasnitz, 2010). Should Assistive Technologies aim for normality or embrace and foster identity, i.e., the pride of being a person with a disability? Culture also mediates expectations we have for social robots in home and other everyday contexts, which is a rising trend (Breazeal et al., 2016). Should social robots maintain interdependence with other members of the household, which is a key trait of collectivist cultures (Hofstede et al., 2010; Triandis, 1995), or perform tasks independently, i.e., without direct human intervention? (Lee et al., 2012)

Freehand gestures we might make to interact with objects in a virtual reality environment are not the same across the world. (Wu et al., 2020) found that cutting throat was the gesture for rejecting a call among US participants while the top gesture for Chinese ones was waving back and forth. Non-Western perspectives on togetherness, orality and storytelling, community based moderation, relation to land, reciprocity, and collectivism are often marginalized in mainstream ICTs (REF-Interact_noticeboard). Yet, by and large, HCI has not kept pace with changes in the user base. To illustrate this fact briefly, if we regard CHI as the flagship of the HCI community, HCI is largely WEIRD (Western, Educated, Industrialized, Rich, and Democratic): between 2016 and 2020, at least 73% of CHI study findings were based on Western participants (Sturm et al., 2015). Although there has been, and continues to be, a strong bias towards WEIRD users, a growing number of studies and efforts argue for 'moving the center' towards cultural diversity (Ai He et al., 2015; Arawjo, 2020; Cannanure, Nemer & Sturm, 2022). There is no 'global' user (Visescu, 2021). Also, these users are no longer passive elements in a human–machine system; they are autonomous individuals with values, goals and beliefs about life and work (Bannon, 1991).

3.2.3   Usability Is Not the Same Here and There

Much research assumes that usability is understood similarly across cultures (Frandsen-Thorlacius et al., 2009). However, cultural differences in understanding the concept of usability have been found. For example, Frandsen-Thorlacius et al. (2009) shows that Chinese users place more emphasis on visual appearance, satisfaction, and fun than Danish people, who place more importance on effectiveness and the lack of frustration. Winschiers and Fendler (2007) argues that the widely presumed correlation between user satisfaction and efficient and effective task completion does not hold for Namibians. For them, usability means beneficial, transparent, flexible, and goal-oriented. Implicit Western understanding of usability might not hold to other cultural groups. This diversity in the meaning of usability suggests that the extent to which the results from usability studies can (and if so, how should) be compared when people with different cultural backgrounds fill them out is fraught with uncertainty.

Usability evaluation methods are based on the premise that participants find it easy to articulate their thoughts and feel comfortable to say what works for them and what does not. However, these assumptions are debatable. As stated in Chavan (2005), one of the most difficult problems in evaluating any technology in India is that its (national) culture largely advocates acceptance of the state of a given situation and then if possible to work around it. To give an obviously negative opinion about people or things is uncomfortable for most people. Also, getting an individual opinion is difficult because India is a collectivist culture and opinions are always collective in nature. To work around these cultural issues, an evaluation 'Bollywood style' has been put forward. Bollywood evaluation draws on, and makes the most of, an accepted critiquing format in India (film reviews).

Clemmensen (2011), Vatrapu and Perez-Quinones (2006) reinforce the need for understanding the impact of culture on usability methods and adapting methods to different cultures. Clemmensen (2011) examines how usability testing methods are applied in industry in Beijing, Mumbai, and Copenhagen. Three distinct different ways of performing a usability test are found: evaluator-centered (Beijing), user-centered (Mumbai), and client-centered (Copenhagen). Beijing is evaluator-centered because the evaluators, who are usually highly qualified human factors professionals, chose the methodology in each case of a usability test. In Mumbai, the task scenario is influenced by where geographically the test takes place. The moderator's skill level is related to the user's personality and test methodology. Verbalization depends on the user's motivation and is considered during test protocol development. The overall user-evaluator relationship is related to the user's motivation and is considered during test protocol development. Finally, Copenhagen is client-centered because there is a strong focus on business goals. The context of success criteria (e.g., increased sales) is a relevant part of the usability test that only the Danish informant mentioned. Using clickable prototypes is therefore an important part of usability testing. Vatrapu and Perez-Quinones (2006) had two groups of participants, all graduate students from India, evaluate a website. One group had an interviewer from India. The other group had an interviewer from the United States of America. The group with the Indian interviewer found more errors, provided more feedback about the website, and identified more culturally-sensitive materials than the group with the US interviewer.

These results show that differences in performing usability evaluation methods exist and that these are not merely a subjective issue in the sense of being random or arbitrary (Clemmensen, 2011). Rather, these differences can be dealt with and explained by considering culture. Hence, usability might not be a definitive (i.e., universal) concept. Hertzum in his article "Images of Usability" (2010) examines usability from a sensitizing point of view. Hertzum (2010) argues that existing approaches to usability form different images of usability, which in spite of a shared essence display profound differences in focus, scope, mindset, and perspective. The shared essence of the images focuses on the common-sense meaning of usability as indicating that a thing is fit, convenient, or ready for use and often also implies that use is experienced as easy and intuitive. The different images include *universal usability*, which entails embracing the challenge of making systems for everybody to use, *situational usability*, which refers to the quality-in-use of a system in a specific situation with its users, tasks, and wider context of use, and *cultural usability*, which stresses the fact that usability takes on different meanings depending on the users' cultural background. Cultural usability emphasizes adaptation (of the meaning of usability and how it is conducted) for usability studies to be meaningful (Douglas & Liu, 2011).

3.2.4 Culture Matters in Technology Design

Winschiers (2006) argues that it has to be recognized that Participatory Design (PD) in a cross-cultural context goes beyond the involvement of users in the design of the product but should include an appropriation of the design process itself. Participatory interactions in oral cultures (e.g., Africa) rely on verbal and performed actions rather than paper or technology-based artifacts. Participation is also a long-term established practice in these cultures. Thus, the emphasis is no longer on facilitating a joint design activity, which brings individuals together and is a key aspect of 'Western' PD, but rather on guiding a closed group towards a design output (Winschiers-Theophilus et al., 2010). To do so, it is important to feel 'being participated', i.e., an uncomfortable sensation (for most Westerner PD practitioners) for the loss of design process control that is followed by a feeling of release that the community is empowered to lead their own process though in a different way.

The cultural question in PD goes beyond the whole way of doing and appropriating PD (Irani, 2010). It also involves the notion of *cultural hybridity* (Merritt & Stolterman, 2012), i.e., seeing culture as a single, unified 'thing' or thinking of it as a multifaceted entity (Burke, 2009; Hakken & Maté, 2014). Merritt and Stolterman (2012) challenges the static cultural binary opposition between the self and the other, Western and non-Western, or the designer and the user, offering a more nuanced approach to understanding the malleable nature of culture. For example, a Western modernizing agenda is often viewed by those in developing countries as inherently imposing cultural changes on intended users. Thus, PD practitioners are encouraged to explore the history of cultural interactions and identities of potential users, and bear in mind notions of power, as being present as an outsider with means and resources for generating a technological intervention implies a position of dominance that accompanies a designer's cultural identity. Hakken and Maté (2014) argues against PD's tendency to conceive of culture as a single, unified 'thing'. Cultural hybridity encourages us to think in the plural and acknowledge diversity (Burke, 2009). Yet, Hakken and Maté (2014) argues that cultural hybridity seems to address culture as a unified phenomenon. However, culture can be seen as a *complex* set of analytic constructs (e.g., values, practices). Hakken and Maté (2014) encourages us to deal with culture in PD by narrowing attention to one aspect of culture at a time and finally assembling them into a single account. For example, Qi2He is a co-design framework inspired by Eastern epistemology (Bryan-Kinns et al., 2022). It explores how we can develop methods to support engagement in cross-cultural co-design activities. The authors adapted traditional Chinese literature composition methods to structure our co-design method. The key features are that there should be stages of technical skill development and sharing, cultural sharing, co-design innovation and evaluation across cultures using in-situ making, and co-design refinement, production, presentation, and evaluation.

There is also a need to provide HCI designers with practical guidance on how to explicitly integrate culture in the design process. Internationalization (i.e., isolating the culturally specific elements from a product) and localization (i.e., infusing a specific cultural context into an internationalized product) is the most widely used cross-cultural design approach (Salgado, Faria, & Sieckenius de Souza, 2013; Singh & Pereira, 2005). However, it assumes national cultural determinism (see Chap. 2) and fails to consider diversity. For example, some users might have more exposure to English language than English speakers do to non-English websites, and this difference in exposure might imply that they do not find it so detrimental to interact with websites with different cultural attributes than those of their national culture (Alexander et al., 2021). Also, it separates design from use (REF-Aykin), and the resultant technological products are detached from their context(s) and meaning(s), which is of paramount importance from a cultural perspective (Marsden, Maunder & Parker, 2008; Bourges-Waldegg & Scrivener, 1998). In a user interface, it is not enough to understand the words. A user interface is acted on, not simply read (REF-Aykin). Ethnography is also regarded by some as instrumental in designing more culturally-sensitive technologies (REFs). However, this approach emphasizes a degree of cultural uniqueness—thick description and authenticity—and this may make it seem impossible to bridge cultural differences (Aykin, 2005), i.e., ending up designing N interfaces, one for each cultural group (or subgroup). Tools, approaches, and toolkits have been proposed to help designers take culture into consideration throughout the design of interactive systems. Some of them are summarized next.

For example, Salgado et al. (2013) introduces Cultural Viewpoint Metaphors (CVM), a Semiotic Engineering conceptual tool aimed to help HCI designers organize culture-sensitive interaction discourse about the journeys that users may take. Pereira and Baranauskas (2015) puts forward a value-oriented and culturally informed approach to design that offers artifacts and methods to address values and culture in a theoretically founded and explicit way. VCIA has three key principles: (1) values and culture are intertwined and inseparable, (2) designers need practical artifacts and methods to support their activities, and (3) designers need a well-designed design process. While VCIA Pereira and Baranauskas (2015) is a designer-generated framework, Alshehri, Kirkham & Olivier (2020) puts forward a culturally-sensitive method designed to elicit values from people who may be in a power relationship that hinders directly discussing implicit values. This method has three elements: Scenarios, Co-creation, and Cards. Hao et al. (2017) focuses on empathy, which is important in design but difficult to achieve when the designer and the target user do not share the cultural context, and introduces *Cultura,* a communication toolkit that uses nine cultural aspects based on cultural models so as to inform designers about user insights in a broader cultural context. The nine cultural aspects include socio-cultural values, the material world, community, division of roles, rituals, knowing the rules, angels and devils, goals of end-users, and macro developments.

Design guidelines also play an important role in operationalizing culture in HCI design. Marcus and Gould (2000) introduces dimensions of culture, as analyzed by Hofstede et al (2010), and considers how they might affect user interface design. Marcus and Gould (2000) shows examples from the Web that illustrate each cultural dimension and also puts forward a number of user interface and web design aspects that might be influenced by them. For example, Power Distance might impact hierarchies, tall vs. shallow, and the prominence given to leaders versus citizens, customers, or employees. Individualism and Collectivism might influence the importance given to individuals or groups, as well as the motivation based on personal or group achievements. Studies show the correlation between these guidelines and cultural models and user interface design. For example, Alexander, Murray & Thompson (2017a, 2017b) conducts an audit of 460 websites (Australian, Chinese, and Saudi Arabian), finding that a heavy use of images of leaders, elderly individuals as well as political and religious images can be related to the high-power distance and collectivistic culture of the Chinese and Saudi Arabian cultures. Images of people in daily life, nature, and images that promote values characteristic of individualist societies are more common in Australian websites. Research is currently being conducted to develop cross-cultural web design guidelines that include but go beyond Hofstede's cultural dimensions. This research is prompted by previous work claiming that prior cross-cultural design guidelines are not sensitive enough, usually considered limited website design attributes, and lack usability tests to support their claims (Alexander et al., 2017a, 2017b).

Cultural adaptation is another example of the importance of culture in design. Cultural adaptation takes design guidelines and cultural dimensions at its core and goes one step further: automatic adaptation. Reinecke and Bernstein (2011), Singh and Pereira (2005) argue that with the Internet being an increasingly global marketplace, it is unreasonable to design one common website for everyone, and yet expect to attract an international audience. A more desirable solution would be to design many different versions of a web page in order to cater to users' cultural preferences. Working towards this goal, the approach 'cultural adaptivity' is proposed, which consists of designing systems that automatically adapt themselves to the user's cultural background. Recabarren and Nussbaum (2010) adopts this approach to automatically adapt webforms to the culture (nation) of end-users, finding that the time taken to fill our web forms can be reduced by cultural adaptation. Leiva and Alabau (2015) acknowledges that the need to modify an application so that it can support different languages and cultural settings can appear once the application is finished and even in the market, which may introduce time delays and an increase in costs. To deal with this problem, the authors introduce a post-hoc method to automatically internationalize websites and web-based applications, without having to modify the source code.

Overall, this overview shows that culture touches upon relevant aspects related to technology design, from ways of doing and appropriating PD and tools for enabling designers to operationalize the concept of culture in design to strategies for systematizing the design of more culturally-sensitive user interfaces.

3.2.5 An Endless Loop: Culture in Technology, Technology in Culture

Culture and technology take part in an endless loop (du Gay et al., 2003). At one end of the loop, culture in technology stresses the fact that technologies are cultural artifacts. Technologies are laden with human, cultural, and social values (Harper et al., 2007). Technology creators, by them designers or programmers, inscribe their views of users and use in technological objects (Chien et al., 2020; Hodgson et al., 2013; Oudshoorn & Pinch, 2003) and algorithms (Airoldi, 2022). Sometimes, these values are anticipated and designed for. For example, the designs of most complex systems (e.g., plane cockpits) are based on the concepts and cultures of low PDI, high IDV, and low UAI of Anglo and North European countries. As a result, the operational environments and training programs associated with these sociotechnical systems incorporate inbuilt cultural assumptions (e.g., the rapid flow of factual information between all team members) which clash with the national cultures of many operators (Chien et al., 2020; Hodgson et al., 2013). Most of the technologies designed for older people draw upon stereotyped (mostly negative) views of older people and aging (Sayago, 2019). We tend to consider that aging is a problem and technologies the solution. We live (especially in the West) in the culture of the youth (Cordella & Poiani, 2021); what is old is to be avoided or disregarded. The widespread application of Artificial Intelligence (AI) and algorithms have led to significant concerns about how modern AI systems are amplifying and perpetuating inequity and privilege in society (Benjamin, 2019; Umoja, 2018). These systems and algorithms are mostly designed (and trained) by young White people (Airoldi, 2022; Cave & Dihal, 2020). On other occasions, however, human, cultural, and social values emerge in interaction. For example, as much as mobile phones are pieces of technology, there are also constellations of social and cultural practice (Bell, 2006). Mobile phones are objects for communication, a source of information, and sites of anxiety and control, and also say something about how their owners are. Emojis are becoming an increasingly popular mode of communication between individuals worldwide. Yet, Kimura-Thollander and Kumar (2019) reveals that "heavy Western tech bias" of the Unicode standard for emojis does not cater well enough to cultural diversity. People from different cultural backgrounds feel that there are emojis missing, mostly related to food and words, and phrases in their native language. This gap is important because emojis are seen as powerful in allowing us to express our cultural identity.

Technology in culture is the other end of the loop and refers to the extent to which technologies shape our everyday lives. Technology and algorithms are more than tools. For example, the Japanese lunchbox is a technology that can be regarded as an emblem of the Japanese way of making things, e.g., compact, waste-avoiding culture, eye-catching, flexibility (Ekuan, 2000). Most of us keep technologies like old mobile phones or cars that do not work at all but tell something about our identities and our past, and evoke (positive or negative) emotions (Norman, 2004). Social media technologies have become part of everyday social practices and altered our experience of sociality. Likes, dislikes, and rankings of popularity are rooted in an ideology that values hierarchy and competition (Van Dijck, 2013). Technology enables us to explore life online and simulate different identities (Turkle, 1995). It also provides us with ever-expanding social connections; yet, it leaves us increasingly disconnected from themselves and unsure of how to make authentic communication with others (Turkle, 2011). Technologies also have the potential to both erode and preserve culture and identity (Goulding et al., 2020), due to issues of power, i.e., who controls and designs the technology (Airoldi, 2022). Algorithms (and others, such as programming) are altering how culture has long been practiced. The sorting, classifying, and characterizing of people, places, objects and ideas are increasingly being delegated to algorithms (Dourish, 2016; Striphas, 2015). This affects, for example, the cultural contents (for example, films or songs) we consume online, as the ones with fewer clicks or less related to our interests might be rendered invisible (Baeza-Yates, 2018). The potential cohabitation between humans and social (or sociable) robots, which are designed to interact with people in a natural, interpersonal manner (Breazeal et al., 2016), is facilitating the emergence of the robotics culture, wherein culture might no longer be a by-product of humans (Samani et al., 2013).

The endless loop outlined in this section shows that technology and culture are difficult to separate from each other. This interconnection is important for HCI to develop a richer understanding of the different ways in which technologies are designed and used, and the consequences (and meanings) of technology use in our different ways of living.

3.2.6 HCI Evolution and Challenges

The field of HCI has evolved over the years. We no longer view users only as a set of (most cognitive) characteristics (Bannon, 1991) and human cognition as dissociated from or unaltered by socio-cultural contexts (Card et al., 1983; Hutchins, 1995). The area of concern is now much broader than the initial 'fit' of people and technologies to improve productivity and usability. It also encompasses new or different forms of living with and through digital technologies (Bannon, 1991, 2011; Bødker, 2006; Grudin, 2012, 2017; Harrison et al., 2011; Hassenzahl, 2010). This evolution comes with a number of challenges. Shneiderman et al. (2016) identify a set of grand challenges, from developing a handbook of human needs and supporting successful aging strategies to encouraging reflection and mindfulness, and promoting lifelong learning. Addressing these, and other challenges, which have been outlined in previous sections in this chapter, needs to consider the cultural factor. In fact, culture is regarded as a key ingredient of the current (or third) wave/paradigm of HCI research (Bødker, 2006; Harrison et al., 2011), which aims to examine, understand and design the aforementioned 'new or different forms of living with and through technologies'.

References

Ai He, H., Memarovic, N., Sabiescu, A., & De Moor, A. (2015). CulTech2015: Cultural diversity and technology design. C&T, 27–30-June, 153–156. https://doi.org/10.1145/2768545.2768561.

Airoldi. (2022). *Machine habitus. Toward a sociology of algorithms.* Polity.

Alexander, R., Murray, D., & Thompson, N. (2017a). Cross-cultural web design guidelines. In *Proceedings of the 14th Web for All Conference*, W4A 2017. https://doi.org/10.1145/3058555.305 8574.

Alexander, R., Thompson, N., & Murray, D. (2017b). Towards cultural translation of websites: A large-scale study of Australian, Chinese, and Saudi Arabian design preferences. *Behaviour & Information Technology, 36*(4), 351–363. https://doi.org/10.1080/0144929X.2016.1234646.

Alexander, R., Thompson, N., McGill, T., & Murray, D. (2021). The influence of user culture on website usability. *International Journal of Human-Computer Studies, 154*, 102688. https://doi.org/10.1016/j.ijhcs.2021.102688.

Alshehri, T., Kirkham, R., & Olivier, P. (2020). Scenario co-creation cards: A culturally sensitive tool for eliciting values. In *Conference on Human Factors in Computing Systems—Proceedings* (Issue 1, pp. 1–14). https://doi.org/10.1145/3313831.3376608.

Arawjo, I. (2020). To write code: The cultural fabrication of programming notation and practice. In *Conference on Human Factors in Computing Systems—Proceedings* (pp. 1–15). https://doi.org/10.1145/3313831.3376731.

Aykin, N. (Ed.). (2005). *Usability and internationalization of information technology.* Lawrence Erlbaum Associates Publishers. https://doi.org/10.1109/memb.2006.1578651

Baeza-Yates, R. (2018). Bias on the web. *Commun. ACM 61*, 6(June 2018), 54–61. https://doi.org/10.1145/3209581.

Bannon, L. J. (1991). From human factors to human actors the role of psychology and human-computer interaction studies in systems design. In *Design at work: Cooperative design of computer systems* (pp. 25–44).

Bannon, L. (2011). Reimagining HCI: Toward a more human-centered perspective. *Interactions, 29*, 50–57.

Bell, G. (2006). The age of the thumb: A cultural reading of mobile technologies from Asia. *Knowledge, Technology, & Policy, 19*(2), 41–57. https://doi.org/10.14361/9783839404034-004.

Benjamin, R (2019). *Race after technology. Abolitionist tools for the new Jim code.* Polity.

Bødker, S. (2006). When second wave HCI meets third wave challenges. NordiCHI, October, 14–18.

Bourges-Waldegg and Scrivener, 1998 Bourges-Waldegg, P., & Scrivener S. A. R. (1998). Meaning, the central issue in cross-cultural HCI design. Interacting with Computers, 9(3), 287–309. https://doi.org/10.1016/S0953-5438(97)00032-5

Breazeal, C., Dautenhahn, K., & Kanda, T. (2016). Social robotics. In B. Siciliano & O. Khatib (Eds.), *Springer handbook of robotics* (pp. 1935–1961). Springer.

Bryan-Kinns, N., Wang, W., & Ji, T. (2022). Qi2He: A co-design framework inspired by eastern epistemology. *International Journal of Human-Computer Studies, 160*, 102773. https://doi.org/10.1016/j.ijhcs.2022.102773.

Burke, P. (2009). *Cultural hybridity.* Polity Press.

Cannanure, V. K., Nemer, D., & Sturm, C. (2022). HCI across borders: Navigating shifting borders at CHI. In *Conference on Human Factors in Computing Systems—Proceedings*.

Card, S., Moran, T., & Newell, A. (1983). *The psychology of human-computer interaction.* Lawrence Erlbaum Associates.

Cave, S., & Dihal, K. (2020). The whiteness of AI. *Philosophy & Technology, 33*, 685–703. https://doi.org/10.1007/s13347-020-00415-6.

Chavan, A. L. (2005). Another culture, another method evaluation bollywood style. *Human Factors*.

Chien, S.-Y., Lewis, M., Sycara, K., Kumru, A., & Liu, J.-S. (2020). Influence of culture, transparency, trust, and degree of automation on automation use. *IEEE Transactions on Human-Machine Systems, 50*(3), 205–214. https://doi.org/10.1109/THMS.2019.2931755.

Clemmensen, T. (2011). Templates for cross-cultural and culturally specific usability testing: Results from field studies and ethnographic interviewing in three countries. *International Journal of Human-Computer Interaction, 27*(7), 634–669. https://doi.org/10.1080/10447318.2011.555303.

Clemmensen, T., & Roese, K. (2010). An overview of a decade of journal publications about culture and Human-Computer Interaction (HCI). *IFIP Advances in Information and Communication Technology, 316*, 98–112. https://doi.org/10.1007/978-3-642-11762-6_9.

Cordella, M., & Poiani, A. (2021). *Fulfilling ageing*. Springer.

Del Galdo, E. M., & Nielsen, J. (Eds.). (1996). *International user interfaces*. Wiley.

Douglas, I., & Liu, Z. (Eds.). (2011). *Global usability*. Springer, London. https://doi.org/10.1007/978-0-85729-304-6.

Dourish, P. (2016). Algorithms and their others: Algorithmic culture in context. *Big Data and Society, 3*(2), 1–11. https://doi.org/10.1177/2053951716665128.

du Gay, P., Hall, S., Janes, L., Mackay, H., & Negus, K. (2003). *Doing cultural studies. The story of the Sony Walkman*. Sage.

Ekuan, K. (2000). *The aesthetics of the Japanese lunchbox*. The MIT Press.

Fernandes, T. (1994). Global interface design: A guide to designing international user interfaces. In *CHI* (pp. 373–374).

Frandsen-Thorlacius, O., Hornbæk, K., Hertzum, M., & Clemmensen, T. (2009). Non-universal usability? A survey of how usability is understood by Chinese and Danish users. In *Conference on Human Factors in Computing Systems—Proceedings* (pp. 41–50). https://doi.org/10.1145/1518701.1518708.

Friedman, T. (2005). *The world if flat. A brief history of the 21st century*.

Goulding, A., Campbell-Meier, J., & Sylvester, A. (2020). Indigenous cultural sustainability in a digital world: Two case studies from Aotearoa New Zealand. In A. Sundqvist, G. Berget, J. Nolin, & K. I. Skjerdingstad (Eds.), *Sustainable digital communities* (Vol. 12051, pp. 66–75). Springer International Publishing. https://doi.org/10.1007/978-3-030-43687-2_5.

Grudin, J. (2012). A moving target: The evolution of human-computer interaction. In J. Jacko (Ed.), *The human-computer interaction handbook*. Fundamentals, evolving technologies, and emerging applications. CRC Press.

Grudin, J. (2017). From tool to partner: The evolution of human-computer interaction. In *Synthesis lectures on human-centered informatics* (Vol. 10, No. 1). Morgan and Claypool. https://doi.org/10.2200/S00745ED1V01Y201612HCI035.

Hakken, D., & Maté, P. (2014). The culture question in participatory design. In *Proceedings of the 13th Participatory Design Conference: Short Papers, Industry Cases, Workshop Descriptions, Doctoral Consortium Papers, and Keynote Abstracts* (Vol. 2, pp. 87–91). https://doi.org/10.1145/2662155.2662197.

Hao, C., Van Boeijen, A., & Stappers, P. J. (2017). Cultura: A communication toolkit for designers to gain empathic insights across cultural boundaries. In *Proceedings of the IASDR Conference RE: Research* (pp. 497–510). https://doi.org/10.7945/C2SD5J.

Harper, R., Rodden, T., Rogers, Y., & Sellen, A. (Eds.). (2007). *Being human: Human computer interaction in the year 2020*. https://doi.org/10.1145/1467247.1467265.

Harrison, S., Sengers, P., & Tatar, D. (2011). Making epistemological trouble: Third- paradigm HCI as successor science. *Interacting with Computers, 23*(5), 385–392. https://doi.org/10.1016/j.intcom.2011.03.005.

Hassenzahl, M. (2010). *Experience design*. Springer.

Hertzum, M. (2010). Images of usability. *International Journal of Human-Computer Interaction, 26*(6), 567–600. https://doi.org/10.1080/10447311003781300.

Hodgson, A., Siemieniuch, C. E., & Hubbard, E.-M. (2013). Culture and the safety of complex automated sociotechnical systems. *IEEE Transactions on Human-Machine Systems, 43*(6), 608–619. https://doi.org/10.1109/THMS.2013.2285048.

Hofstede, G., Hofstede, G. J., & Minkov, M. (2010). *Cultures and organizations.* McGraw Hill.

Hutchins, E. (1995). *Cognition in the wild.* The MIT Press.

Irani, L. (2010). HCI on the move: Methods, culture, values. In *Conference on Human Factors in Computing Systems—Proceedings* (pp. 2939–2942). https://doi.org/10.1145/1753846.1753890.

Kamppuri, M., Bednarik, R., & Tukiainen, M. (2006). The expanding focus of HCI: Case culture. *ACM International Conference Proceeding Series, 189*(October), 405–408. https://doi.org/10. 1145/1182475.1182523.

Kimura-Thollander, P., & Kumar, N. (2019). Examining the "global" language of emojis: Designing for cultural representation. In *Conference on Human Factors in Computing Systems—Proceedings* (pp. 1–14). https://doi.org/10.1145/3290605.3300725.

Lee, H. R., Sung, J., Sabanovic, S., & Han, J. (2012). Cultural design of domestic robots: A study of user expectations in Korea and the United States. In *Proceedings—IEEE International Workshop on Robot and Human Interactive Communication* (pp. 803–808). https://doi.org/10.1109/ ROMAN.2012.6343850.

Leiva, L. A., & Alabau, V. (2015). Automatic internationalization for just in time localization of web-based user interfaces. *ACM Transactions on Computer-Human Interaction, 22*(3), 1–32. https:// doi.org/10.1145/2701422.

Linxen, S., Cassau, V., & Sturm, C. (2021). Culture and HCI: A still slowly growing field of research. Findings from a systematic, comparative mapping review. In *ACM International Conference Proceeding Series.* https://doi.org/10.1145/3471391.3471421.

Mankoff, J., Hayes, G. R., & Kasnitz, D. (2010). Disability studies as a source of critical inquiry for the field of assistive technology. In *ASSETS'10—Proceedings of the 12th International ACM SIGACCESS Conference on Computers and Accessibility* (pp. 3–10). https://doi.org/10.1145/187 8803.1878807.

Marcus, A., & Gould, E. W. (2000). Cultural dimensions and global web user-interface design. *Interactions,* 32–46.

Marcus, A. (2001). International and intercultural user interfaces. In Stephanidis, C (Ed.) User interfaces for all. Concepts, Methods, and Tools (pp. 47–65). Lawrence Erlbaum Associates.

Marsden, G., Maunder, A., & Parker, M. (2008). People are people, but technology is not technology. *Philosophical Transactions of the Royal Society A: Mathematical, Physical and Engineering Sciences, 366*(1881), 3795–3804. https://doi.org/10.1098/rsta.2008.0119

Merritt, S., & Stolterman, E. (2012). Cultural hybridity in participatory design. *PDC,* 73–76. https:// doi.org/10.1145/2348144.2348168.

Nielsen, J., Del Galdo, E. M., & Sprung, R. C. (1990). Designing for international use. In *Conference on Human Factors in Computing Systems—Proceedings,* April (pp. 291–294). https://doi.org/10. 1145/97243.97298.

Norman, D. (2004). *Emotional design.* Basic Books.

Oudshoorn, N., & Pinch, T. (Eds.). (2003). *How users matter. The co-construction of users and technologies.* The MIT Press. https://doi.org/10.1353/tech.2006.0041.

Pereira, R., & Baranauskas, M. C. C. (2015). A value-oriented and culturally informed approach to the design of interactive systems. *International Journal of Human Computer Studies, 80,* 66–82. https://doi.org/10.1016/j.ijhcs.2015.04.001.

Recabarren, M., & Nussbaum, M. (2010). Exploring the feasibility of web form adaptation to users' cultural dimension scores. *User Modeling and User-Adapted Interaction, 20*(1), 87–108. https://doi.org/10.1007/s11257-010-9071-7.

Reinecke, K., & Bernstein, A. (2011). Improving performance, perceived usability, and aesthetics with culturally adaptive user interfaces. *ACM Transactions on Computer-Human Interaction, 18*(2), 1–29. https://doi.org/10.1145/1970378.1970382.

Russo, P., & Boor, S. (1993). How fluent is your interface? Designing for international users. *Interchi*, 342–347.

Salgado, L., Faria, C., & Sieckenius de Souza, C. (2013). *A journey through cultures.* Springer Human-Computer Interaction Series.

Samani, H., Saadatian, E., Pang, N., Polydorou, D., Fernando, O. N. N., Nakatsu, R., & Koh, J. T. K. V. (2013). Cultural robotics: The culture of robotics and robotics in culture. *International Journal of Advanced Robotic Systems, 10*, 1–10. https://doi.org/10.5772/57260.

Sayago, S (Ed.) (2019). *Perspectives on human-computer interaction research with older people.* Springer Human-Computer Interaction Series.

Shneiderman, B., Plaisant, C., Cohen, M., Jacobs, S., Elmqvist, N., & Diakopoulos, N. (2016). Grand challenges for HCI researchers. *Interactions, 23*(5), 24–25. https://doi.org/10.1145/2977645.

Singh, N., & Pereira, A. (2005). *The culturally customized web site.* Elsevier.

Striphas, T. (2015). Algorithmic culture. *European Journal of Cultural Studies, 18*(4–5), 395–412. https://doi.org/10.1177/1367549415577392.

Sturm, C., Oh, A., Linxen, S., Abdelnour-Nocera, J., Dray, S., & Reinecke, K. (2015). How WEIRD is HCI? Extending HCI principles to other countries and cultures. In *Conference on Human Factors in Computing Systems—Proceedings* (Vol. 18, pp. 2425–2428). https://doi.org/10.1145/2702613.2702656.

Triandis, H. (1995). *Individualism & collectivism.* Routledge.

Turkle, S. (1995). *Life on the screen. Identity in the age of the Internet.* Simon & Schuster.

Turkle, S. (2011). *Alone together.* Basic Books.

Umoja, S (2018). *Algorithms of oppression: How search engines reinforce racism.* New York University Press.

Van Dijck, J. (2013). *The culture of connectivity.* Oxford University Press.

Vatrapu, R., & Perez-Quinones, M. (2006). Culture and usability evaluation: The effects of culture in structured interviews. *Journal of Usability Studies, 1*(4), 156–170. http://citeseerx.ist.psu.edu/viewdoc/summary?doi=10.1.1.101.5837.

Visescu, I. D. (2021). The impact of culture on visual design perception. In C. Ardito, R. Lanzilotti, A. Malizia, H. Petrie, A. Piccinno, G. Desolda, & K. Inkpen (Eds.), *Human-Computer Interaction—INTERACT 2021* (Vol. 12936, pp. 499–503). Springer International Publishing. https://doi.org/10.1007/978-3-030-85607-6_66.

Winschiers, H. (2006). The challenges of participatory design in an intercultural context: Designing for usability in Namibia. In *Proceedings of the Participatory Design Conference* (pp. 73–76). http://ojs.ruc.dk/index.php/pdc/article/view/375.

Winschiers, H., & Fendler, J. (2007). Assumptions considered harmful the need to redefine usability. In N. Aykin (Ed.), *Usability and Internationalization, Part I, HCII 2007.*

Winschiers-Theophilus, H., Chivuno-Kuria, S., Kapuire, G. K., Bidwell, N. J., & Blake, E. (2010). Being participated—A community approach *PDC*, 1–10. https://doi.org/10.1145/1900441.1900443.

Wu, H., Gai, J., Wang, Y., Liu, J., Qiu, J., Wang, J., & Zhang, X. (2020). Influence of cultural factors on freehand gesture design. *International Journal of Human Computer Studies, 143*(February 2019). https://doi.org/10.1016/j.ijhcs.2020.102502.

You, S., Kim, M., & Lim, Y. (2016). Value of culturally oriented information design. *Universal Access in the Information Society, 15*(3), 369–391. https://doi.org/10.1007/s10209-014-0393-9.

Conceptual Perspectives of Culture Within HCI

This chapter presents conceptual perspectives of culture within HCI. The Cultural Dimensions of Hofstede et al. (2010), the cross-cultural theory of communication of Hall (1973, 1976, 1982), and the cultural cognitive systems of thought of Nisbett (2003), which predominate in much HCI research (especially Cultural Dimensions, see Chap. 5), are summarized. To provide a richer conceptual perspective, this chapter also outlines other views of culture which, despite not playing a central role in HCI, deepen and extend the main ones. It summarizes the work of Trompennars and Hampden-Turner in *Riding the waves of culture*. This work shares with Hosftede the national and dimensions approach. As opposed to Cultural Dimensions, in *Riding the waves of culture* nations are not given a static index in each dimension and it is acknowledged that these dimensions are complementary, not opposing, preferences. It also outlines the work of Triandis on *Individualism/Collectivism* (1995), which addresses a key cultural dimension in other areas (de Mooij, 2011; Heine, 2016; Keith, 2019) as well as deepening one of the dimensions of the Hofstede's model. This chapter also summarizes Swidler's view of culture as a 'toolkit' (1986), from which actors select differing pieces for constructing different lines of action rather than pushing action in a consistent direction. This view regards culture as a sensitizing concept (Blume, cited in Puddephatt et al., 2009, p. 19), as opposed to the definitive concept of culture presented in the aforementioned conceptual perspectives. This chapter also deals with subcultures and counter-cultures, which align with the use of the plural of the term culture and is an example of the cultural diversity of nations, groups of people, and individuals. Table 4.1 presents a summary of the conceptual perspectives outlined.

© The Author(s), under exclusive license to Springer Nature Switzerland AG 2023
S. Sayago, *Cultures in Human-Computer Interaction*, Synthesis Lectures
on Human-Centered Informatics, https://doi.org/10.1007/978-3-031-30243-5_4

Table 4.1 Conceptual perspective of culture in HCI

Conceptual perspective	Summary
Culture as the software of the mind	Culture is the collective programming of the mind that distinguishes the members of one group or category of people from others. Cultural dimensions explain national differences
Culture as communication	Communication constitutes the core of culture. Different types of communication: high-context, low-context, spatial, and proxemics
Culture as a system of thought	People see the world in terms dictated by their social existence. Two cognitive processes: holistic and analytic
Culture as solving problems	Culture is the way in which a group of people solves problems and reconciles dilemmas. Seven fundamental dimensions of culture, i.e., problems that seem universal and each culture addresses in different ways
Individualism versus Collectivism	Individualistic and collectivist patterns operate across cultures and within any society and, in fact, within each human
Culture as a toolkit	Culture is like a 'toolkit' from which actors select differing pieces for constructing lines of action. Two people who have very similar cultural repertoires may nonetheless differ greatly in the emphasis they accord particular views
Subcultures and counter-cultures	Cultural groups interact based on a dominant signifying order but there may be more than one order acting inside. One of these orders can be subcultures, counter-cultures, or forms of resistance

4.1   Main Conceptual Perspectives

4.1.1   Culture as the Software of the Mind

Hofstede, in *Cultures and Organizations. Software of the Mind* (Hofstede et al., 2010), intends to "help in dealing with the differences in thinking, feeling, and acting of people around the globe. It will show that although the variety in people's minds is enormous, there is a structure in this variety that can serve as a basis for mutual understanding" (p. 4). According to Hofstede, culture "is the collective programming of the mind that distinguishes the members of one group or category of people from others" (p. 6). This software of the mind does not mean that a person is unable to react in ways that are new, creative, or unexpected. The software of the mind indicates what reactions are likely and understandable, given one's past (p. 5), since culture is learned, not innate.

Based on statistical analyses of survey data (over 116,000 questionnaires with more than 100 standardized questions each), covering IBM employees in 72 subsidiaries, 38 occupations, 20 languages, and at two points in time, around 1968 and 1972 (Hofstede et al., 2010), the model distinguishes cultures of different nationalities along six dimensions: power distance, individualism/collectivism, masculinity/femininity, uncertainty avoidance, long-/short-term orientation, indulgence/restraint. These dimensions explain, in statistical terms, almost all the differences observed in the survey data.

Power distance (PDI) is the extent to which the less powerful members of institutions and organizations within a country expect and accept that power is distributed unequally. Institutions are the basic elements of society, such as the family, the school, and the community; organizations are the places where people work (p. 61). In small PDI countries (these include the United States, Great Britain, and much of the rest of Europe), there is limited dependence of subordinates on bosses. In large PDI countries (which tend to exist in Latin America and in Asia), there is considerable dependence of subordinates on bosses.

Individualism pertains to societies in which ties between individuals are loose: everyone is expected to look after him- or herself and his or her immediate family. *Collectivism* as its opposite pertains to societies in which people from birth outward are integrated into strong, cohesive in-groups (p. 92) Examples of countries that value individualism include the United States, UK, Netherlands, France, Germany, and Canada. Some values include personal time, freedom, and challenge. Examples of countries that value collectivism include Japan, Costa Rica, Mexico, and Greece. Some values include training, physical strength, and the use of skills.

Examples of countries where the *feminine* index is more valued include Sweden, Israel, and Indonesia. The feminine index includes aspects such as more leisure time is preferred over more money, and people work in order to live (p. 170). Examples of countries where the *masculine* index is more valued include Ireland and Great Britain, Philippines, and Hungary. The masculine index includes aspects such as boys play to compete and girls play to be together, and being responsible, decisive, and ambitious is for men; being caring and gentle is for women.

Uncertainty avoidance is the extent to which the members of a culture feel threatened by ambiguous or unknown situations (p. 191). In countries with weak uncertainty avoidance (e.g., Sweden and Denmark), what is different is curious, whereas in countries with strong uncertainty avoidance, such as Japan and South Korea, what is different is dangerous.

Long-term orientation stands for fostering of virtues oriented toward future rewards. The values for long-term orientation are persistence and perseverance, respect for a hierarchy of the status of relationships, and having a sense of shame. Examples of countries that have a long-term orientation toward life are China, Taiwan, and Japan. It's opposite, *short-term orientation,* stands for the fostering of virtues related to the past and present (p. 239). The values for short-term orientation are a sense of security and stability, protecting your reputation, respect for traditions, and reciprocation of greetings and. Examples of countries that have a short-term orientation toward life include Pakistan, Mexico, and Australia.

Indulgence stands for a tendency to allow relatively free gratification of basic and natural human desires related to enjoying life and having fun. Its opposite pole, *restraint*, reflects a conviction that such gratification needs to be curbed and regulated by strict social norms (p. 281). In high indulgent countries (e.g., Venezuela) there are higher percentages of very happy people and a perception of personal life control. In high restraint countries (e.g., Egypt) there are lower percentages of very happy people and a perception for helplessness: what happens to me is not my own doing.

The model claims that these dimensions explain national cultural differences: "IBMers do not form representative samples from national populations (...) but they are so similar in respects other than nationality (...) that the only thing that can account for systematic and consistent differences between national groups within such a homogenous multinational population is nationality itself (2001, pp. 251–252)".

Hofstede operationalizes these dimensions in *Exploring Cultures. Exercises, stories, and synthetic cultures* (Hofstede, 2002), which "presents training that uses all of Hofstede's dimensions (...) in the development of cross-cultural training" (p. x). The synthetic cultures "bring Hofstede's dimensions of national culture to life in activities and simulations" (p. 1), from unexpected meetings to job interviews. These synthetic cultures are ten, two for five dimensions: *Indiv* for extreme individualism, *Collect* for extreme collectivism; *Hipow* for extremely large power distance, *Lopow* for extremely small power distance; *Mascu* for extreme masculinity, *Femi* for extreme feminity; *Uncavo* for extremely strong uncertainty avoidance, *Unctol* for extreme uncertainty tolerance; *Lotor*, extreme long-term orientation, *Shotor*, extreme short- term orientation.

4.1.2 Culture as Communication

The three works of Eduard T. Hall outlined in this section (*The Silent Language*, *The Hidden Dimension*, and *Beyond Culture*) draw mostly on "years of involvement with the selection and training of Americans working in foreign countries for both government and business" (Hall, 1973, p. ix). In these works, Hall focuses on communication, as "most people's difficulties with each other can be traced to distortions on communications" (Hall, 1973). In particular, Hall concentrates on covert or 'invisible' communication. This type of communication, which he refers to as the silent language or the hidden dimension, because it is taken for granted and "outside voluntary control" (Hall, 1982), is of paramount importance in the interaction between people, since we "'talk to each other without the use of words" (Hall, 1973, p. vii). For Hall, communication constitutes the core of culture.

In *The Silent Language* (Hall, 1973), Hall argues that "just because we talk does not mean the rest of what we communicate with our behavior is not equally important" (p.

viii). Hall argues that *time* and *space* 'talks'. We will refer to space later on, as Hall develops the concept of *space talks* in greater detail in the Hidden Dimension.

The concept of time varies across cultures. According to Hall, for Americans, it is somewhat immoral to have two things going on at the same time, and if people are not prompt, it is often taken as an insult or as an indication that they are not quite responsible. This conceptualization of time is not the same across cultures. Hall talks about polychromic cultures (e.g., most countries in Latin America) wherein two or more than two things can go on at the same time. Yet, most of us overlook the role that time plays in how we transmit messages in our own culture/s and the hidden rules that govern it. Hall encourages us to be aware of our own and others' silent and seemingly irrational language (Hall, 1976) to "reach other people both inside and outside our national boundaries, as we are increasingly required to do" (Hall, 1982).

In the *Hidden Dimension* (Hall, 1982), Hall focuses on *proxemics*, i.e., the interrelated observations and theories of man's use of space. Hall emphasizes that virtually everything that a man is and does is associated with the experience of space, which he defines as a basic, underlying organizational system for all living things—particularly for people. In the case of wild animals, for example, Hall points out that they allow a man or other potential enemy to approach only up to a given distance before it flees (or attacks). Distance, therefore, plays a fundamental role in their survival.

Hall distinguishes four different types of distances in the human species: intimate, personal, social, and public. These types of distances depend on the transaction, the relationship of the interacting individuals, how they feel, and what they are doing.

Intimate distance is the distance of love making and wrestling, comforting, and protecting. Personal distance is the distance separating the members of non-contact species. It might be thought of as a small protective sphere—'you are invading my space'. Social distance is the distance to which people move when someone says, '*Stand away so I can look at you*'. Intimate visual details in the face are not perceived, nobody touches or expects to touch another person, and formal business activities tend to be conducted within this distance. In public distance, we have exaggerated voices and gestures.

Hall provides a number of examples of proxemics in cross-cultural contexts. Spatial and architectural needs are not the same for all of us. According to Hall, in the United States, space is used as a way of classifying people and activities, whereas in the UK it is the social system that determines who you are rather than where you live or work. Space does not mean the same in the West and in the East. In the West, we are taught to perceive, and react to the arrangements of objects, and to think of space, as "empty." The meaning of this becomes clear only when it is contrasted with the Japanese, for example, who are trained to give meaning to spaces. The *ma*, or interval, is a basic building block in Japanese spatial experience. It is functional not only in flower arrangements but apparently, is a hidden consideration in the layout of all other spaces. According to Hall, for the Arabs, there is no such thing as an intrusion in public. Public means public. In fact, in the Middle East, Americans, in Hall's view, can be compressed and overwhelmed by smells, crowding, and high noise levels in the street. Space talks.

In *Beyond culture* (Hall, 1976), Hall goes one step further and encourages us to go beyond ethnocentrism, be open, understand ourselves better, and embrace diversity, because "we must accept the fact that there are many roads to truth" (p. 7). He acknowledges that this is not easy, because culture is an irrational force, "deeply entrenched in the lives of all of us, and because of culturally imposed blinders, our view of the world does not normally transcend the limits imposed by our culture" (p. 219).

Hall distinguishes between two broad types of communication, which tend to be used as a framework in studies of intercultural communication in HCI (see Part II); a *high-context* communication in which most of the information is either in the physical context of internationalized in the person, and a *low-context* communication which is just the opposite. The Japanese culture belongs to high-context communication whereas the US belongs to the low-context communication. Hall also argues that no culture exists exclusively at one end of this scale.

4.1.3 Culture as a System of Thought

Nisbett in *The Geography of Thought* (2003) argues that very different systems of perception and thought exist. By drawing on historical and philosophical evidence, as well as social science research, including ethnographies, surveys, and laboratory research, Nisbett proposes that people see the world in terms dictated by their social existence.

East Asians live in an interdependent world in which the self is part of a larger whole, while Westerners live in a world in which the self is a unitary free agent. Easterners value success and achievement in good part because they reflect well on the groups they belong to, whereas Westerners value these things because they are badges of personal merit.

These social differences affect the nature of cognitive processes. Nisbett and Norenzayan (2001) find East Asians to be *holistic*, attending to the entire field and assigning causality to it, making relatively little use of categories and formal logic, and relying on "dialectical" reasoning, whereas Westerners are more *analytic*, paying attention primarily to the object and the categories to which it belongs and using rules, including formal logic, to understand its behavior.

Consider for example what people see when they look at a scene. In one study, with fascinating results, American and Japanese participants were shown some animated computer images of an underwater scene and asked to describe what they saw. The Japanese participants made about 60% more references to background objects (holistic) than the Americans, who tended to talk more about the fish at the center of the scene (analytic) (cited in Keith, 2019, p. 354). When looking at an identical scene, Americans and Japanese appear to perceive it differently.

Cultural Psychology gives support to Nisbett's claim by arguing (and showing) that the mind cannot be understood without reference to the sociocultural environment to which it is adapted and attuned (Heine, 2016).

4.2   Other Related Perspectives

4.2.1   Culture as Solving Problems

Individualism/Collectivism (I vs. We) is one of the cultural dimensions of Hofstede, Trompennars, and Hampden-Turner. Individualism/Collectivism is also a key cultural dimension in other areas, such as (Cross-) Cultural Psychology (Heine, 2016), and Consumer Behavior, wherein it is suggested that Individualism/Collectivism accounts for most of the cultural differences among people (de Mooij, 2011). Section 4.2.1 addresses this cultural trait. Trompennars and Hampden-Turner in *Riding the waves of culture* (1997) deal with "cultural differences and how they affect the process of doing business and managing" (p. 1). Culture is defined as "the way in which a group of people solves problems and reconciles dilemmas" (p. 6).

As opposed to Hofstede, Trompennars and Hampden-Turner state that culture presents itself on different levels: national, corporate, and professional. Yet, Trompennars and Hampden-Turner focus on culture at a national level, as Hofstede does. All of them acknowledge that the essence of culture is not what is visible on the surface. It is however the shared ways of groups of people understand and interpret the world. Thus, people within a culture do not all have identical sets of norms, values, and assumptions, but share a number of them.

By drawing mostly on 15 years of academic and field research, over 1000 cross-cultural training programs conducted in more than 20 countries, and a questionnaire completed by 30 K participants spanning 50 different countries, Trompennars and Hampden-Turner argue that there is no "one best way of managing" (p. 2), and examine seven fundamental dimensions of culture. In line with their view of culture (i.e., solving problems), these dimensions are problems that seem universal and each culture addresses in different ways. According to Trompennars and Hampden-Turner, people everywhere are confronted with three sources of challenges: relationships with other people, time, and environment. Five dimensions cover the ways in which human beings deal with each other.

Universalism versus particularism. The universalist is rule-based. S/he assumes that the one good way must always be followed. By contrast, the particularist places more attention to the unique circumstances – no matter what the rules say. Will you cross the street when the light is red but no car approaches (no matter what the rules say), or will you wait until the light turns green regardless of the traffic (rule-based)? What do people in your country do?

Individualism versus communitarism (the group versus the individual). We deal with this dimension in Individualism/Collectivism (Sect. 2.3.1).

Neutral versus emotional. In this dimension, we can classify business relationships that are typically instrumental and all about achieving objectives (e.g. British, North Americans), while others in which the human affair and emotions are very important (e.g. Italians).

Specific versus diffuse (the range of involvement). A specific relationship is prescribed by a contract while in a diffuse one, there is a real and personal contact. In specific-oriented cultures, such as France or Germany, "a manager segregates out the task relationship she or he has with a subordinate and insulates this from other dealings" (p. 81).

Achievement versus ascription. This is all about how we accord status. While some societies accord status to people on the basis of their achievements (e.g., the United States), others ascribe it to them by virtue of age, class, gender, education, and so on (e.g., the Czech Republic). The first kind of status is called achieved and the second ascribed status.

The remaining two dimensions deal with attitudes to time and the environment. Regarding time, Trompennars and Hampden-Turner consider past achievements and future plans. They argue that the American Dream (future looking) is the French Nightmare (an enormous sense of the past). With respect to the environment, some cultures see nature as something more powerful than individuals, whereas others see the major focus affecting their lives as residing within the person. For example, with mobile devices that enable us to listen to music, we might decide to use them because 'I can listen to music without being disturbed by other people' or 'I can listen to music without bothering people'. These represent very different attitudes towards the environment.

As opposed to Cultural Dimensions, wherein nations are given a static index in each dimension, Trompennars, and Hampden-Turner consider that "it is probably true to say these dimensions (*the ones in their book*) are complementary, not opposing, preferences. They can each be effectively *reconciled* by an integrative process" (p. 52, *emphasis added*). This dynamism concurs with the work of Triandis on a key dimension: Individualism/Collectivism, which is outlined later.

4.2.2   Individualism/Collectivism

Triandis, in *Individualism/Collectivism* (1995), focuses on this cultural trait. *Individualism/Collectivism* "was written to help the reader understand how the individualistic and collectivist patterns operate. They operate both across cultures and within any society and, in fact, within each human. There is a constant struggle between the collectivist and individualist elements within each human (…) In collectivist cultures people act like collectivists in most situations in which they are dealing with an ingroup, but they act like individualists, maximizing their benefits and outcomes, in most situations where they deal with outgroups" (p. xiv).

Collectivism may be initially defined as a social pattern consisting of closely linked individuals who (i) see themselves as parts of one or more collectives (family, co-workers, tribe, nation), (ii) are primarily motivated by the norms of, and duties imposed by those collectives, (iii) are willing to give priority to the goals of these collectives over their own personal goals, and (iv) emphasize their connectedness to members of these collectives.

Individualism, however, is a social pattern that consists of loosely linked individuals who (i) view themselves as independent of collectives, (ii) are primarily motivated by their own preferences, needs, rights, and the contracts they have established with others, (iii) give priority to their personal goals over the goals of others, and (iv) emphasize rational analyses of the advantages and disadvantages to associating with others (p. 2).

The degree of individualism or collectivism in any given culture is influenced by certain factors (p. 52). Individualism attains its highest levels when I-factors (cultural complexity, affluence, maleness, urbanism, high social class, and social and geographic mobility) are large and C-factors small. Collectivism attains its highest levels when C-factors (cultural homogeneity, high population density, and isolation from other cultures/groups, and they are greater still when there is an external threat and the members of the culture realize that survival depends on interdependence) are large and I-factors are small (p. 81).

In addition to these factors, cultural *tightness/looseness*, and cultural *complexity/simplicity*, stand out in Individualism/Collectivism. Individualism is most often a consequence of looseness and cultural complexity, while collectivism is most often a consequence of tightness and cultural simplicity (p. 52). Both tightness and looseness are situation-specific. A culture may be tight in social and political situations and loose in economic or religious situations. Thus, when we state that a culture is tight (or loose), we mean that tightness (or looseness) is characteristic across many, but not all, situations (p. 53). Culture emerges in interaction (p. 4).

4.2.3 Culture as a Toolkit

While Hofstede's Cultural Dimensions (2010), and to some extent the cross-cultural theory of communication of Hall (1973, 1976, 1982) and the cultural systems of thought of Nisbett (2003), seem to assume that culture pushes action in a consistent direction, e.g., if you are from the East you belong to the high-context communication and analytic thought group, and your behavior can be predicted and classified according to some dimensions, Swidler (1986) argues that culture is more like a 'toolkit' from which actors select differing pieces for constructing lines of action.

We make choices about what cultural meanings to accept and how to interpret them, and two people who have very similar cultural repertoires may nonetheless differ greatly in the emphasis they accord particular views. This point is illustrated in her study of love among a relatively homogenous group of middle-class Americans, who drew from a common pool of cultural resources. What differentiates them is how they make use of the culture they have available. Two distinct cultures of love were found. When thinking about the choice of whether to marry or stay married, people see love in mythic terms. Love is the choice of one right person. When thinking about maintaining ongoing relationships, people mobilize the prosaic-realistic culture of love to understand the varied ways one can manage love relationships (Swidler, 2001).

Swidler also argues that the way culture influences action differs in two kinds of situations: settled and unsettled (Swidler, 1986). The contrast is intended to different situations in which new strategies of action are being developed and tried out (unsettled) from situations in which people are operating within established strategies of action (settled). In unsettled lives, such as those lived by most adolescents, culture is more visible because people actively use culture to learn new ways of being.

While Hofstede' Cultural Dimensions, the cross-cultural theory of communication of Hall and the cultural systems of thought of Nisbett allows us to understand how one culture differs from another, Swidler argues that the main analytic problem should be to describe the varied ways people use diverse cultural materials, appropriating some and using them to build a life, holding others in reserve, and keeping still others permanently at a distance (Swidler, 1986, 2001).

4.2.4   Subcultures/Counter-Cultures

While culture is often associated with nation, a noteworthy example being the Cultural Dimensions proposed by Hofstede, there is widespread agreement that most nations have several cultures, and that most individuals integrate multiple cultural identities (Smith-Jackson, Resnick & Johnson, 2014). Cultural groups interact based on a dominant signifying order but there may be more than one order acting inside (Salgado, Faria & Sieckenius de Souza, 2013). One of these orders can be subcultures, counter-cultures, or forms of resistance, i.e., expressive forms of groups who are often oppressed, treated as second-class citizens or deviants (Hebdige, 1979).

Culture is for example both a source of oppression and of liberation for disabled people (Riddell & Watson, 2014). On the one hand, the most predominant cultural construction of old age and disability (at least, in Western countries) promotes negative views of both groups, presenting them as deficient categories, in opposition to an idealized notion of adulthood, failing to acknowledge the salience of social exclusion in marginalizing them (Riddell & Watson, 2014). On the other hand, people with disabilities have forged their own identities as a form of resistance, i.e., the Disability culture (Peters, 2000).

In terms of identity, disability, and culture, the Deaf culture is worth noting. *Deaf* people are those hearing-impaired people—with their national/regional culture—who use sign language and distance themselves from both phonocentric world and any suggestion that they are people with impairments or disabled people—sign language is the 'natural' language of Deaf people (Bragg et al., 2019).

References

Bragg, D., Koller, O., Bellard, M., Berke, L., Boudreault, P., Braffort, A., Caselli, N., Huenerfauth, M., Kacorri, H., Verhoef, T., Vogler, C., & Morris, M. R. (2019). Sign language recognition, generation, and translation: An interdisciplinary perspective. In *ASSETS 2019—21st International ACM SIGACCESS Conference on Computers and Accessibility* (pp. 16–31). https://doi.org/10. 1145/3308561.3353774.

de Mooij, M. (2011). *Consumer behavior and culture. Consequences for global marketing and advertising.* Sage.

Hall, E. (1973). *The silent language.* Anchor Books.

Hall, E. (1976). *Beyond culture.* Anchor Books.

Hall, E. (1982). *The hidden dimension.* Anchor Books.

Hebdige, D. (1979). *Subculture.* Routledge.

Heine, S. (2016). *Cultural psychology* (3rd ed.). W. W. Norton & Company.

Hofstede, G. (2002). *Exploring culture. Exercises, stories, and synthetic cultures.* Intercultural Press.

Hofstede, G., Hofstede, G. J., & Minkov, M. (2010). *Cultures and organizations.* McGraw Hill.

Keith, K. (Ed.). (2019). *Cross-cultural psychology. Contemporary themes and perspectives.* Wiley Blackwell.

Nisbett, R. (2003). The geography of thought. *The Free Press.* https://doi.org/10.1111/j.0033-0124. 1973.00331.x

Nisbett, R. E., & Norenzayan, A. (2001). Culture and systems of thought: Holistic versus analytic cognition. *Psychological Review, 108*(2), 291–310.

Peters, S. (2000). Is there a disability culture?: A syncretisation of three possible world views. *Overcoming Disabling Barriers: 18 Years of Disability and Society, 15*(4), 583–601. https://doi.org/ 10.4324/9780203965030.

Puddephatt, A., Shaffir, W., & Kleinknecht, S. (Ed.). (2009). *Ethnographies revisited. Constructing theory in the field.* Routledge.

Riddell, S., & Watson, N. (Eds.). (2014). *Disability, culture and identity.* Routledge.

Salgado, L., Faria, C., & Sieckenius de Souza, C. (2013). *A journey through cultures.* Springer Human-Computer Interaction Series.

Smith-Jackson, T. L., Resnick, M. L., & Johnson, K. (Eds.). (2014). *Cultural ergonomics. Theory, methods, and applications.* CRC Press.

Swidler, A. (1986). Culture in action: Symbols and strategies. *American Sociological Review, 51*(2), 273–286.

Swidler, A. (2001). *Talk of love.* University of Chicago Press. https://doi.org/10.7208/chicago/978 0226230665.001.0001

Triandis, H. (1995). *Individualism & collectivism.* Routledge.

Trompenaars, F., & Hampden-Turner, C. (1997). *Riding the waves of culture. Understanding cultural diversity in business.* Nicholas Brealey Publishing.

The Operationalization of Culture in HCI

5

This chapter aims to provide an overview of how culture is operationalized in HCI research. It presents a profile of research that identifies particular traits of HCI research wherein culture is a primary topic of interest. It summarizes two main research approaches to study culture in this field and presents examples of HCI studies adopting them. This chapter also outlines the main functions of culture in HCI research.

5.1 A Profile of Research: Increasing, Mostly Quantitative

The culture-related studies[1] published during several periods (1990–2005; 2010, and 2016–2020) in major HCI journals and ACM CHI have been analyzed (Kamppuri et al., 2006; Clemmensen & Roese, 2010; Linxen et al., 2021). This section brings the individual analysis together to provide an integrated profile of research. Table 5.1 summarizes the results.

5.1.1 Period: 1990–2005. Sources: Journals and ACM CHI

In 2006, a quantitative content analysis of culture-related studies published in four HCI journals (Human-Computer Interaction, International Journal of Human-Computer Interaction, International Journal of Human-Computer Studies, and Interacting with Computers) and ACM CHI conference between 1990 and 2005 was published (The expanding focus of HCI: Case culture). The results show that HCI articles considering national/ethnic

---

[1] Studies with "culture" or "cultural" in the title, keywords, or abstract.

© The Author(s), under exclusive license to Springer Nature Switzerland AG 2023
S. Sayago, *Cultures in Human-Computer Interaction*, Synthesis Lectures
on Human-Centered Informatics, https://doi.org/10.1007/978-3-031-30243-5_5

Table 5.1 A profile of HCI research on culture

Analysis	Summary of findings
From 1990 to 2005. Journals and ACM CHI (Kamppuri et al., 2006)	Cultural HCI is comparative (and mostly quantitative) and based on traditional human factors studies. Culture as a characteristic of a user. Nearly 40% of the articles left culture without any definition, and in those where culture was defined, the most common source was Hofstede's Cultural Dimensions.
From 1998 to 2007. Journals (Clemmensen & Roese, 2010)	Most publications reported quantitative research in which the participants were university students that spoke English. Almost half of the studies did not use any theory or model at all. Most studies are cross-cultural and use national groups as cultural groups.
From 2016 to 2020 (2010 included too). Journals and ACM CHI (Linxen et al., 2021)	Hofstede dominates the theoretical frameworks with one third of the papers using his definition. Quantitative methods predominate, with an increase in qualitative methods. The four most common national cultures considered are US-American, Chinese, British, and South Korean.

culture[2] were rare in the first half of the 1990s; yet, there was a noticeable increase (from 4 to 24 articles published) during the second half of the 16-year period. The results also show that the Internet and groupware were the most common technological contexts in the studies, which applied mostly comparative studies using questionnaires, formal experiments, and interviews. The results indicate that most of the studies (57%) considered culture as a characteristic of a user, i.e., culture affecting their cognitive style, and that the most common national cultures studied were American, followed by Chinese, Japanese, and German. Nearly 40% of the articles left culture without any definition, and in those where culture was defined, the most common source was Hofstede's Cultural Dimensions.

In summary, (Kamppuri et al., 2006) points out that the prevailing methodology in cultural HCI is comparative and based on traditional human factors studies. The underlying cultural theory is borrowed mostly from the studies of cultural dimensions that are often considered controversial (see Chap. 6), and more efforts should be spent on studying the cultures from within. To do so, (Kamppuri et al., 2006) argues for employing more contextual, ethnographic studies, to provide new viewpoints and complement traditional approaches.

---

[2] The number of articles in the final simple, after removing articles that did not meet the inclusion criteria, was 28.

5.1.2   Period: 1998–2007. Source: Journals

(Clemmensen and Roese, 2010) provides an overview of publications in culture and HCI between 1998 and 2007 with a narrow focus on journal publications (full papers) only. Seven journals were considered (Interacting with Computers, International Journal of Industrial Ergonomics, Behaviour and Information Technology, International Journal of Human-Computer Interaction, Computers in Human Behaviour, International Journal of Human-Computer Studies, Journal of Usability Studies) and the total number of papers included in the analysis was 27. The results show a small, but continuous stream of cultural usability HCI journal papers published during the time period considered, and that most of these papers (20) reported quantitative research in which the participants were (mostly) university students that spoke English. The results show that Hofstede's Cultural Dimensions were used by 7 of the 27 studies, and that almost half of studies (12) did not use any theory or model at all. The results indicate that the most common research approach was experimental (19 of 27), with China and the US as the anchor country in most of the studies, which were largely cross-cultural and used national groups as cultural groups.

5.1.3   Period: 2010; 2016–2020. Sources: Journals and ACM CHI

(Linxen et al., 2021) follows up on (Kamppuri et al., 2006) by conducting a systematic quantitative content analysis of full articles published in seven high profile journals and conference proceedings in 2010, and from 2016 to 2020. The journals were the same as those used in (Kamppuri et al., 2006) plus two more, Behaviour and Information Technology and ToCHI. The results show that the number of articles (N = 109) dealing with culture increased steadily, especially from 2016 to 2020. Following (Kamppuri et al., 2006), the articles were classified according to their technological areas, finding that nearly 40% addressed general design issues, while groupware declined considerably (5.5%). The results also indicate that half of the papers used quantitative research methods, and that studies that considered the cultural context of a user were most common than in (Kamppuri et al., 2006). The results show that the four most common national cultures were US-American, Chinese, British, and South Korean. Hofstede dominates the theoretical frameworks with one third of the papers using his definition of culture as "software of the mind", while a large number of studies (50) did not refer to any specific concept of the theory of culture.

Overall, (Linxen et al., 2021) shows that HCI community has developed an increased but moderate interest in cultural topics; quantitative methods are still used predominantly, although an increase in qualitative methods might have been expected to explore in-depth culture. (Linxen et al., 2021) also shows that theoretical sources are still not included in a large part of the papers, which report research conducted in a small number of nations.

(Linxen et al., 2021) encourages the HCI community to further increase diversity and address culture more systematically.

5.2   Two Main Approaches: Taxonomic and Contingent

(Halabi and Zimmermann, 2019) argues that HCI design (and in fact, research too) dealing with culture can be seen from two main different approaches/perspectives[3]: the taxonomic and the contingent. The *taxonomic perspective* renders human ways of living and thinking into a finite set of elements sorted into categories. The *contingent perspective,* however, looks at culture as a dynamic, constructed, and interactional phenomenon.

This section presents HCI studies which adopt these approaches in an attempt to give a flavor of each of them. Further examples can be found in Appendix I. HCI studies that adopt the taxonomic perspective (see Sect. 5.2.1) often draw on the Cultural Dimensions of Hofstede, the cross-cultural theory of communication of Hall (low and high context), and the cultural cognitive systems of thought (analytic versus holistic) of Nisbett. These studies tend to be quantitative and most of them are experimental. This perspective predominates in much HCI research. The studies dealing with cultural adaptation, design dimensions, and guidelines outlined in Chap. 3 (Sect. 3.2.3) fall under this perspective. The contingent perspective (Sect. 5.2.2) is not so widely adopted. Studies addressing the culture question in Participatory Design and in most of the studies dealing with usability (Chap. 3, Sects. 3.2.2 and 3.2.3) embrace it. HCI studies that adopt this perspective tend to be qualitative, most of them rely on a combination of first-hand observations and conversations, and are conducted in out-of-laboratory settings.

5.2.1   Examples of Taxonomic Research

Computer-Mediated Communication: The increasing globalization of the workplace, home life, and education has invigorated communication and collaboration among people from a variety of geographical locations around the world, usually via computer-mediated communication (CMC) technologies (Setlock & Fussell, 2011). The taxonomic perspective is common within studies of CMC. Some studies are summarized next.

---

[3] Another approach might be related to the one adopted in (intangible) cultural heritage. Most of the studies in this area that have been reviewed for this synthesis, including the proceedings of Cross-Cultural Design (Rau, 2018, 2020) do not fit in either taxonomic or contingent perspectives, as they tend to focus on the design, evaluation and / or development of novel technological systems to augment and personalize the cultural heritage experience in some way (Lu et al., 2011; Wecker, Kuflik & Stock, 2017; Cardoso et al., 2020; Avram & Maye, 2016; Volkmar, Wenig & Malaka, 2018; Lu et al., 2019; Ruiz-Calleja et al., 2023; Karran et al., 2015; Gou, 2021; Petrelli, 2019; Not & Petrelli, 2019; Ardissono et al., 2012). This perspective is mostly technological and the cultural aspect is narrowed down to museum and arts.

(Aragon and Poon, 2010) conducted a case study of the deployment of context-linked software tools (where both task and context information are directly included in the shared communication space) in international astrophysics collaboration (between the US and France), and described examples where these tools facilitated cross-cultural communication in the framework of Hoftede's cultural dimensions. Although there have been many critiques of Hofstede's work, there are few frameworks available to use to study cultural differences and discuss cross-cultural collaboration (p. 159). The results found aspects of cross-cultural communication that demonstrated cultural differences according to the Cultural Dimensions (e.g., French participants stated in an interview that they did not usually ask questions on the shift, expecting to the told what to do by the experts; this is typical of a high power distance society), and also that context-linked software tools fostered the bridging of these differences in cultural styles over time.

(Nguyen and Fussell, 2012) examines whether or not people encounter more problems working with a partner from a different culture than with one from the same culture by focusing on Instant Messaging conversations between Americans and Chinese. These two groups have been shown to prefer distinct styles of communication. American speakers tend to be direct and to the point (low context), whereas Chinese speakers tend to be more indirect, relying on context to make their messages clear (high context). The results of an experimental study show that Chinese participants reported more problems than American participants regardless of the culture of their partners, and that participants, regardless of their own cultures, reported higher annoyance working with a Chinese partner–probably, because high-context communicators can adjust flexibility to match the styles of low context communicators.

Web page perception: (Baughan & Reinecke, 2021) and (Dong & Lee, 2008) are two studies that illustrate important elements of the taxonomic perspective, such as cultural models, in research on web page perception.

(Baughan and Reinecke, 2021) draw on Hosftede's Cultural Dimension and the cross-cultural theory of communication to Hall. The authors report on an online study with US American and Japanese participants (mean age 33 and 29.6, respectively), asking them to search for specific information on a set of website screenshots of varying complexity. The study aimed to example whether the participants attended to other parts of a website while engaged in the primary search task. US Americans and Japanese were selected because Japanese have repeatedly been found to focus on contextual information more than US Americans. The results of the online study did not confirm that Japanese participants were faster at finding and better at recalling, contextual information than participants from the US. Yet, Japanese took three times as long to find information on highly complex websites than US Americans, and this might account for the process of familiarizing with a website being different between Japanese (holistic) and US American (analytic) participants.

Based on Nisbett's cognitive model of holistic and analytic thought, (Dong & Lee, 2008) hypothesizes that differences between holistic thought and analytic thought can be reflected in webpage perception. The results of an eye-tracking experimental study in which American, Chinese, and Korean participants (between the ages of 24 and 25) were asked to look at a webpage (the version corresponding to their national culture) without clicking on anything as the task was to determine how people actually view a webpage and thus to reveal their natural viewing pattern, indicate that the Chinese and Korean subjects showed more similarities to holistic thought patterns (i.e., scan back and forth among the page contents), while the American subjects showed more similarities to analytic thought patterns (e.g., read the prototype in sequential order).

Social media: The growth and usage level of social network sites have become a global phenomenon (Vasalou, Joinson & Courvoisier, 2010; Annamoradnejad et al., 2019), and this has stimulated research on the differences and similarities in the way that people in different countries use social media. Some studies that illustrate a key element of the taxonomic perspective (culture = nation) are outlined below.

Does national culture determine the temporal randomness with which Twitter users post, or the extent to which they mention, follow, recommend and befriend others? By analyzing the tweets of 2.34 million user profiles during 10 weeks, (Garcia-Gavilanes, Quercia & Jaimes, 2013) shows that countries with a higher pace of life (e.g., Germany, Switzerland) tend to be more predictable not only offline but also online. Users in monochronic countries tend to be temporarily predictable, those in collectivist countries considerably talk with each other, and those in countries uncomfortable with power distance will not preferentially engage only with popular users.

(Sheldon, Herzfeldt and Rauschnabel, 2020) provides insights into the relationship between cultural values and hashtagging behavior. Hashtags have become an integral part of social media communication. However, our knowledge of how people use hashtags is limited. The results of an online survey with over 200 respondents (average age: 32) show that inspirational hashtags are common among users with collectivistic, uncertainty avoidant, and masculine cultural values. Moreover, collectivistic and masculine values are also associated with artistic hashtags, whereas uncertainty avoidance is related to entertaining hashtags. Findings show that cultural values associated with power distance are related to a higher hashtagging intensity.

(AlMuhanna et al., 2022) address the extent to which the fear of antisocial behavior on Twitter impacts the acceptance and use of that platform. By drawing on an experimental study conducted with two different cultures, Anglophone countries (UK, USA, and Canada), and Non-western Arabic culture of Saudi Arabia, the authors found that users from Anglophone countries who perceive a problem with antisocial behavior on Twitter are not less willing to use it. However, if these users felt that the usefulness of Twitter was being constrained or limited, this would discourage them from using it. The Saudi sample was also negatively affected by the interaction of the perception of antisocial behavior; however, the results showed less impact compared to the Anglophone sample. Individualism and collectivism might explain these differences.

Persuasive technology: (Khaled et al., 2006a, 2006b; Orji & Mandryk, 2014) argues that most persuasive technology strategies cater to a large individualist audience. Following this, (Khaled et al., 2006a, 2006b) proposed five collectivist-focused persuasive strategies–group opinion, group surveillance, deviation monitoring, disapproval conditioning, and group customization. A game titled *Smoke?* about smoking cessation was designed by following these principles. One version was designed to be more persuasive for NZ European (individualist) players, while the other was designed to be more persuasive for Maori (collectivist) players. The results of qualitative evaluation (through focus groups) with students (18-35 years old) support the claim that culturally matched strategies yield greater persuasion (Khaled et al., 2006a, 2006b; Khaled, 2008). (Orji & Mandryk, 2014) focuses on healthy eating behavior, puts forward guidelines for tailoring persuasive technology interventions to collectivists and individualist cultures based on a large-scale survey, and argues that for PT interventions to achieve the objective of promoting healthy eating behavior, they must be culturally relevant, as the widespread 'one-size-fits-all' is heavily biased towards individualist groups.

Interacting with virtual agents: As the development of virtual agents focuses increasingly more on the social aspects of human interaction, it becomes crucial to address the notion of culture and how it affects human behavior. Much of this research adopts a taxonomic perspective.

(Mascarenhas et al., 2016) addresses the challenge of creating virtual agents that are able to portray culturally appropriate behavior when interacting with other agents or humans. Through the use of the Social Importance Dynamics model, this study demonstrates how general cultural tendencies that are shown in everyday social interaction can be explicitly represented in a structured manner, rather than implicitly in the agents' goals and actions. One of the main advantages of making culture explicit in an agent model is that it becomes possible to conduct agent-based simulations and experiments where one manipulates the culture of a group of agents as a separate parameter. Using an interactive narrative scenario that is part of an agent-based tool for intercultural training, this study conducted a cross-cultural study in which participants from a collectivistic country (Portugal) were compared with participants from an individualistic country (the Netherlands) in the way they perceived and interacted with agents whose behavior was either individualistic or collectivistic, according to the configuration of the proposed model. Portuguese subjects rated the collectivistic agents more positively than the Dutch but both countries had a similarly positive opinion about the individualistic agents.

(Aljaroodi et al., 2022) argues that research on avatars and UI design has centered primarily on users from North America and Europe, leaving aside users from various Arab countries. This study aims to address the question of how Cultural appropriateness in avatar design affects users' trusting beliefs and usage intentions. Online experiment with 313 Saudi participants. The experiment is framed in the context of an online health advice service, which is a common application domain for avatars. The findings emphasize that users find avatars that reflect cultural markers from their own background (such as clothes) more culturally appropriate and exhibit higher trust and usage intention toward such avatars.

To increase the level of physical activity among Chinese immigrants in the U.S, especially first generation immigrants whose levels of physical activity are the lowest, (Zhou, Zhang & Bickmore, 2017) reports on an embodied conversational agent to promote physical activity among young Chinese adults living in the U.S. This study adopted Hofstede's cultural dimension theory to design two versions (Chinese and American) of the virtual agents. Participants aged 19 to 34 years old participated in a between-subject pilot study in which they interacted with the agents. The results show that participants correctly identified the Chinese adapted agent to be more Chinese than neutral. This proved that the cultural adaptations in both systems were successfully recognized. The authors hypothesized that the Chinese agent would be more effective in persuading Chinese participants, but instead found a trending effect supporting the opposite.

5.2.2   Examples of Contingent Research

The contingent perspective plays a pivotal role in "postcolonial computing", a project of understanding how all design research and practice is culturally located and power laden, even if considered fairly general (Irani et al., 2010). Rather than classifying people on various cultural dimensions, a more productive analytical position, (Irani et al., 2010) argue, is a *generative* view of culture in which we ask how the technological objects and knowledge practices of everyday life become meaningful. It is precisely those changing cultural practices that designers aspire to support and in which they wish to intervene when they introduce a system into a setting. The studies that follow show examples of this generative view in several areas of interest.

(Jamil et al., 2017) and (Kahn & Burrell, 2021) deal with technology use with very different types of users. (Jamil et al., 2017) focuses on children while (Kahn & Burrell, 2021) concentrates on rural migrants. Both draw on qualitative research and present how technology use and culture unfold.

(Jamil et al., 2017) presents a study of children engaged in a peer-learning task around interactive tabletops in three different countries: the UK, India, and Finland. The study aimed to improve our understanding of the impact of nationality differences on the usage of advanced technology like tabletops in educational settings, since research on this topic is orientated towards Western users. Video was used to record the physical and verbal behavior of all the groups performing the activities in real-world settings, e.g., in the classroom. The results highlight that children in India displayed the most in terms of all the observed physical strategies (positioning, simultaneous object movement and physical contact). Children in the UK exhibited static positioning with some occurrences of simultaneous object movement and physical contact. Meanwhile, children in Finland displayed greater spatial positioning than in the UK with some occurrences of simultaneous object movement and physical contact. Individualism and collectivism might account for these results. The smaller the IDV number, the more the participants will perceive themselves as part of a group (collectivist, India), and the higher the IDV number the more the participants will perceive themselves as individuals (individualist, the UK).

(Kahn and Burrell, 2021) focused on second generation mingongs' experiences with technologies. Mingongs are rural-urban migrants in China. They were either born in the cities their parents worked, or had moved there with their parents at a very young age. They know little to nothing of their parents' rural hometowns and many tend to consider cities as their homes. However, most of them face the same social, cultural, and economical marginalization and exclusion that their parents did. By drawing on 3-month participant observations and semi-structured interviews, the results revealed a new type of "rurality". While living in the cities for almost their entire lifetime, second generation mingongs' experiences with technologies differed in almost every aspect from their urban-native peers. However, they shared some similarities with their peers in rural areas. For those mingongs, interactions with technologies mostly fulfilled routine tasks and leisure purposes but were seldom directed toward self-improvement. Such rurality mostly resulted from their identity struggles. While rejecting urban identity and failing to create their own identities, their nostalgia and memory about their rural hometowns helped them reconstitute rural identities in their interactions with technologies.

(Lyu and Carroll, 2022), (Lindtner, Anderson & Dourish, 2012), (Shao, 2021), and (Prabhakar, Maris & Medhi Thies, 2021) show the extent to which culture determines people's attitudes towards and practices of technology use. (Lyu and Carroll, 2022) addresses the question of how Chinese citizens adopt digital contact tracing in the COVID-19 pandemic, and (Shao, 2021) explores whether and how the unique historical and cultural background of China influence citizens' judgment towards rumor in the face of large-scale epidemic (in this case, COVID-19). (Lindtner, Anderson and Dourish, 2012) examines technology appropriation as a cultural process. (Prabhakar, Maris and Medhi Thies, 2021) presents a preliminary investigation into the cultural influences on social media use of middle class mothers in India. All of them report on qualitative research and show how cultural norms, identities, and values shape technology use.

How did Chinese citizens adopt digital contact tracing in the COVID-19 pandemic? (Lyu & Carroll, 2022) addresses this question by drawing on a series of interviews with Chinese participants who both resided in the mainland of China in the pandemic and used digital contact tracing technologies, i.e., an ICT approach for controlling public health crises. The results indicate that culture (in particular, Confucianism, which is one of the most influential cultures in East Asia) played a significant role in shaping citizens' attitudes toward and practices with digital contact tracing. Most citizens worked collectively as a human infrastructure to support the contact tracing infrastructure. The collectivistic sense motivated citizens to share the responsibility of containing the pandemic with the government. Interviewees also showed an awareness of thinking for the long term. Most participants experienced uncomfortable treatment from local agents. However, such as uncomfortable or even harsh treatment did not lead to resistance due to the influence of Confucianism.

(Shao, 2021) explores whether and how the unique historical and cultural background of China Influence citizens' judgment towards rumor in the face of large-scale epidemic (in this case, COVID-19). Since the outbreak of COVID-19 in 2020, rumors have become increasingly preventable with the popularity of SNS. What is (not) a rumor online? By drawing on semi-structured interviews with active social media users aged between 20 and 22 in China, the results indicate that the participants relied on their existing value systems to guide their attitudes and behaviors towards information when faced with uncertainty. In particular, traditional culture represented by traditional Chinese medicine had a significant influence on the attitudes and behaviors of Chinese citizens, especially those who have benefited from traditional Chinese medicine. Based on these results, (Shao, 2021) proposes the concept of cultural heuristic to consider how, in this case, cultural factors affect Chinese citizens' response to rumors spread on social media.

(Lindtner, Anderson and Dourish, 2012) examines technology appropriation as a cultural process. This study argues that appropriation is not only a matter of unexpected use or customization, but also about the ways that people adapt and make the technology their own. In appropriating technologies, people produce not only content or tweak software code, but also create new meanings and values. The ways in which people appropriate information systems reflects ideas about how they are and who they might be, about how they are connected to others, their roles in relation to them, and how the places we find ourselves in are connected to others. This study grounds these claims in ethnographic research on collaborations and exchanges among IT professionals in urban China.

(Prabhakar, Maris and Medhi Thies, 2021) presents a preliminary investigation into the cultural influences on social media use of middle class mothers in India. This study argues that a great deal of HCI research conducted in India is focused on the low- income population, and that this can result in a particular representation of India in Western dominated research. To fill this gap, this study extends the research space to middle class mothers. Based on semi-structured interviews about social media preferences, online and offline support sources during pregnancy and transition to motherhood, and the influence of cultural and familial norms and expectations of social media participation, the results indicate that cultural and familial traditions and norms influenced social media engagement patterns of the participants. The study found the offline strong-tie relationships of the middle class mother in India, wherein pregnancy and postpartum care are given much emphasis in all strata of Indian society, contributed to low dependence on social media for these issues. Participants reported that they used Facebook to build weak-tie networks and access novel information on pregnancy and parenting (i.e., lurkers), as they relied on WhatsApp, which served as an important extension of their offline strong-tie networks.

The contingent perspective can also be found in technology design studies. For example, (Korte, Potter & Nielsen, 2017) discusses the impacts of Deaf culture and the traits of individual Deafness on the conduct of co-design with Deaf children. Lower case deaf is often used to describe an individual with some form of hearing loss. Upper case Deaf describes individuals who identify as belonging to the Deaf culture. Sign languages are a central component of Deaf cultures, their role in Deaf communities is even characterized as sacred (Bragg et al., 2019). Communication within Deaf culture, being primarily signed, has different cues from communication in hearing culture. Initiating communication requires that visual attention be gained. Common techniques include waving within a person's field of vision, lightly touching their arm or shoulder, or tapping or stamping a surface to cause vibrations. Visual attention must then be maintained throughout the communication. A series of design sessions conducted with young Deaf children of hearing parents indicated that these traits turned out to be a difficult adjustment for hearing people (the designers) to make, and that getting familiar with the local Deaf community was important to keep communicating with them.

5.3   Main Functions

This section outlines the "work" the term culture is doing as it is used in HCI. By drawing on the overview of HCI research on culture presented in this chapter (and in previous ones), this section argues that culture is used in HCI mostly as a way of (a) understanding and finding meaning, (b) explaining differences, (c) dealing with and reducing complexity and (d) a source of inspiration and reflection. Each function is presented individually for comprehensibility purposes. Yet, they are related to each other. In particular, all of them are related to meaning, since developing insights into why people interact with technologies in the way they do, and the importance of doing so, enables us to account for differences, manage complexity, come up with ideas to design better technologies and understand ourselves (and others) better.

5.3.1   Meaning

Culture helps us understand and account for the ways in which people interact with digital technologies as well as their attitudes towards them. For example, Chapter 3 shows that culture mediates the expectations people have for people with disabilities and digital technologies (Mankoff et al., 2010) and the role of social robots in a home in other everyday contexts (Breazeal, Dautenhahn & Kanda, 2016). What a mobile phone is (that is, the meaning of this technology) does not have a straightforward answer. Mobile phones are objects for communication, a source of information, sites of anxiety and control, and also say something about how their owners are (Bell, 2006). Both taxonomic and contingent perspectives show that culture is used as a unit of analysis (e.g., how the behavior and practices of people, such as coordination and collaboration, varies depending on cultural – mostly national – characteristics). In the contingent perspective, culture becomes the vehicle for exploring and understanding people's interactions with technology. Common to both perspectives is the concept of culture as the main determinant of interaction. It is culture, both visible and non-visible manifestations of it (e.g., assumptions, beliefs, values, ways of doing, and thinking) that drive and account for the diverse ways in which we use technologies.

5.3.2 Explain Differences

Culture is also used to summarize the ways in which groups of people distinguish themselves from other groups. For example, lower case deaf is often used to describe an individual with some form of hearing loss. Upper case Deaf describes individuals who identify as belonging to the Deaf culture (Ladd, 2003; Bragg et al., 2019). Retrogamers could be defined as a specific subculture of digital gaming typically characterized by playing (or collecting) obsolete personal computers, consoles, or arcade video games in contemporary times (Mora-Cantallops et al., 2021). Studies that adopt a taxonomic approach are mostly comparative, and virtually all of them report differences. Do we schedule events online (such as videoconferencing meetings) in the same way? (Reinecke et al., 2013) Do all people debug[4] in the same way? (Thayer, Guo & Reinecke, 2018) Do all people participate in Questions and Answers online communities in the same way? (Kayes et al., 2015) Do people with different cultural backgrounds perceive a virtual agent in the same way? (Zhou, Zhang & Bickmore, 2017) The answer to these and other related and similar questions is no, and culture plays a key role in explaining why, i.e., the differences. The overview of studies on usability and culture provided in Chap. 3 shows that differences in understanding and performing usability are not merely a subjective issue in the sense of being random or arbitrary (Clemmensen, 2011). Rather, these differences can be dealt with and explained by considering culture. The studies that adopt a contingent perspective also contribute to using culture as a way of explaining differences by revealing and explaining what is particular to a group of people, and this knowledge helps us understand how different (or similar) other groups can be.

5.3.3 Deal With and Reduce Complexity

Culture is also used to reduce the complexity of people's interactions with digital technologies. For example, we can say that a person uses a technology X in a particular way because s/he belongs to an individualist or a collectivist culture. Yet, one can be individualist or collectivist in a myriad of ways (Triandis, 1995). Internationalization (i.e., isolating the culturally specific elements from a product) and localization (i.e., infusing a specific cultural context into an internationalized product) are two further examples of this use

---

[4] To remove mistakes from a computer program

of culture. As stated in Chap. 3, this approach is common in cross-cultural design (Salgado, Faria & Sieckenius de Souza, 2013). Yet, it separates design from use, e.g., color X should be used to convey a particular meaning when the technological product is designed for country X. This is a way of reducing complexity, as assumes cultural uniformity and detaches the resultant technological products from their interactional context(s) (Dourish, 2004) and consequently from their meanings, which is of paramount importance from a cultural perspective (Marsden, Maunder & Parker, 2008; Bourges-Waldegg & Scrivener, 1998). Design guidelines and cultural adaptation based on Hofstede's Cultural Dimensions are two further examples of this use of culture. Depending on the cultural (national) background of a person, a user interface is designed in a particular way (e.g., different colors and organization). Although this approach is effective (e.g., Recabarren & Nussbaum, 2010), it assumes that culture equals nations, which, as outlined in Chap. 1, is a simplification. A user interface should reflect, among other aspects, the values, ethics, and language of the nation in question (Russo & Boor, 1993). This is a complex issue.

5.3.4 A Source of Inspiration and Reflection

Another function of culture has to do with inspiration and reflection. For example, the design tools, approaches, and toolkits outlined in Chap. 3 can be seen through this inspirational lens. Cultura (Hao, Van Boeijen & Stappers, 2017) aspires to become an inspiring motivation for designers, giving them a broader view of the cultural context in order to build an empathic understanding of the intended users, and go beyond stereotypes. VCIA aims to help designers identify and organize requirements related to the values and culture of the different stakeholders involved in the design context (Pereira & Baranauskas, 2015). Qi2HE aims to support engagement in cross-cultural co-design activities by adapting traditional Chinese literature composition methods to the context of participatory design (Bryan-Kinns et al., 2022). Differences in understanding and performing usability have encouraged the HCI community to adapt usability evaluation methods according to the cultural profile of the participants (see Chap. 3). Research adopting either taxonomic or contingent perspectives also shows that people do not use or perceive the technology in the same way, which challenges widespread assumptions, such as members of different cultures using technology in the same way as designers do (Vatrapu, 2010). Culture reinforces the need to go beyond ethnocentrism, understand others (and ourselves) better, and embrace diversity (Hall, 1976).

References

Aljaroodi, H. M., Adam, M. T. P., Teubner, T., & Chiong, R. (2022). Understanding the importance of cultural appropriateness for user interface design: An avatar study. *ACM Transactions on Computer-Human Interaction, 29*(6), 1–27. https://doi.org/10.1145/3517138

AlMuhanna, N., Hall, W., & Millard D. E. (2022). Fear of the dark: A cross-cultural study into how perceptions of antisocial behaviour impact the acceptance and use of twitter. *Behaviour & Information Technology*, 1–14 https://doi.org/10.1080/0144929X.2022.2064766

Annamoradnejad, I., Fazli, M., Habibi, J., & Tavakoli, S. (2019). Cross-cultural studies using social networks data. *IEEE Transactions on Computational Social Systems, 6*(4), 627–636. https://doi.org/10.1109/TCSS.2019.2919666

Aragon, C. R., & Poon, S. (2010). No sense of distance: Improving cross-cultural communication with context-linked software tools. In *iConference* (pp. 159–165). https://doi.org/10.1145/1940761.1940783

Ardissono, L., Kuflik, T., & Petrelli, D. (2012). Personalization in cultural heritage: The road travelled and the one ahead. *User Modeling and User-Adapted Interaction, 22*(1–2), 73–99. https://doi.org/10.1007/s11257-011-9104-x

Avram, G., & Maye, L. (2016). Co-designing encounters with digital cultural heritage. In *DIS 2016 Companion - Proceedings of the 2016 ACM Conference on Designing Interactive Systems: Fuse* (pp. 17–20). https://doi.org/10.1145/2908805.2908810

Baughan, A., & Reinecke, K. (2021). Do cross-cultural differences in visual attention patterns affect Search Efficiency on Websites ?. In *Conference on Human Factors in Computing Systems - Proceedings*.

Bell, G. (2006). The age of the thumb: A cultural reading of mobile technologies from Asia. *Knowledge, Technology, & Policy, 19*(2), 41–57. https://doi.org/10.14361/9783839404034-004

Bourges-Waldegg and Scrivener, 1998 Bourges-Waldegg, P., & Scrivener S. A. R. (1998). Meaning, the central issue in cross-cultural HCI design. Interacting with Computers, 9(3), 287–309. https://doi.org/10.1016/S0953-5438(97)00032-5

Bragg, D., Koller, O., Bellard, M., Berke, L., Boudreault, P., Braffort, A., Caselli, N., Huenerfauth, M., Kacorri, H., Verhoef, T., Vogler, C., & Morris, M. R. (2019). Sign language recognition, generation, and translation: An interdisciplinary perspective. In *ASSETS 2019-21st International ACM SIGACCESS Conference on Computers and Accessibility* (pp. 16–31). https://doi.org/10.1145/3308561.3353774

Breazeal, C., Dautenhahn, K., & Kanda, T. (2016). Social Robotics. In B. Siciliano & O. Khatib (Eds.), *Springer Handbook of Robotics* (pp. 1935–1961). Springer.

Bryan-Kinns, N., Wang, W., & Ji T. (2022). Qi2He: A co-design framework inspired by eastern epistemology. *International Journal of Human-Computer Studies*, 160, 102773. https://doi.org/10.1016/j.ijhcs.2022.102773

Cardoso, P. J. S., Rodrigues, J. M. F., Pereira, J., Nogin, S., Lessa, J., Ramos, C. M. Q., Bajireanu, R., Gomes, M., & Bica, P. (2020). Cultural heritage visits supported on visitors' preferences and mobile devices. *Universal Access in the Information Society, 19*(3), 499–513. https://doi.org/10.1007/s10209-019-00657-y

Clemmensen, T. (2011). Templates for cross-cultural and culturally specific usability testing: Results from field studies and ethnographic interviewing in three countries. *International Journal of Human-Computer Interaction, 27*(7), 634–669. https://doi.org/10.1080/10447318.2011.555303

Clemmensen, T., & Roese, K. (2010). An overview of a decade of journal publications about culture and human-computer interaction (HCI). *IFIP Advances in Information and Communication Technology, 316*, 98–112. https://doi.org/10.1007/978-3-642-11762-6_9

Dong, Y., & Lee, K. P. (2008). A cross-cultural comparative study of users' perceptions of a webpage: With a focus on the cognitive styles of Chinese, Koreans and Americans. *International Journal of Design, 2*(2), 19–30.

Dourish, P. (2004). What we talk about when we talk about context. *Personal and Ubiquitous Computing, 8*(1), 19–30. https://doi.org/10.1007/s00779-003-0253-8

Garcia-Gavilanes, R., Quercia, D., & Jaimes, A. (2013). Cultural dimensions in twitter: Time, individualism and power. In *Proceedings of the 7th International Conference on Weblogs and Social Media, ICWSM 2013* (pp. 195–204).

Gou, Y. (2021). Computer digital technology in the design of intangible cultural heritage protection platform. In *2021 3rd International Conference on Artificial Intelligence and Advanced Manufacture* (pp. 1524–1528). https://doi.org/10.1145/3495018.3495434

Halabi, A., & Zimmermann, B. (2019). Waves and forms: Constructing the cultural in design. *AI and Society, 34*(3), 403–417. https://doi.org/10.1007/s00146-017-0713-8

Hall, E. (1976). *Beyond Culture.* Anchor Books.

Hao, C., Van Boeijen, A., & Stappers P. J. (2017). Cultura: A communication toolkit for designers to gain empathic insights across cultural boundaries. In *Proceedings of the IASDR Conference RE: Research* (pp. 497–510). https://doi.org/10.7945/C2SD5J

Irani, L., Vertesi, J., Dourish, P., Philip, K., & Grinter, R. E. (2010). Postcolonial computing: A lens on design and development. *Conference on Human Factors in Computing Systems-Proceedings, 2,* 1311–1320. https://doi.org/10.1145/1753326.1753522

Jamil, I., Montero, C. S., Perry, M., O'Hara, K., Karnik, A., Pihlainen, K., Marshall, M. T., Jha, S., Gupta, S., & Subramanian, S. (2017). Collaborating around digital tabletops: Children's physical strategies from India, the UK and Finland. *ACM Transactions on Computer-Human Interaction, 24*(3), 1–30. https://doi.org/10.1145/3058551

Kahn, Z., & Burrell, J. (2021). A sociocultural explanation of internet-enabled work in rural regions. *ACM Transactions on Computer-Human Interaction, 28*(3), 1–22. https://doi.org/10.1145/3443705

Kamppuri, M., Bednarik, R., & Tukiainen, M. (2006). The expanding focus of HCI: Case culture. *ACM International Conference Proceeding Series, 189*(October), 405–408. https://doi.org/10.1145/1182475.1182523

Karran, A. J., Fairclough, S. H., & Gilleade, K. (2015). A Framework for psychophysiological classification within a cultural heritage context using interest. *ACM Transactions on Computer-Human Interaction, 21*(6), 1–19. https://doi.org/10.1145/2687925

Kayes, I., Kourtellis, N., Quercia, D., Iamnitchi, A., & Bonchi, F. (2015). Cultures in community question answering. In *HT 2015-Proceedings of the 26th ACM Conference on Hypertext and Social Media* (pp. 175–184). https://doi.org/10.1145/2700171.2791034

Khaled, R., Barr, P., Fischer, R., Noble, J., & Biddle, R. (2006). Factoring culture into the design of a persuasive game. *OZCHI, 2006*(206), 213–220. https://doi.org/10.1145/1228175.1228213

Khaled, R., Biddle, R., Noble, J., Barr, P., & Fischer, R. (2006b). Persuasive interaction for collectivist cultures. In *Conferences in Research and Practice in Information Technology Series* (pp. 63–70).

Khaled, R. (2008). *Culturally-relevant persuasive technology.* Victoria University of Wellington.

Korte, J., Potter, L. E., & Nielsen, S. (2017). The impacts of deaf culture on designing with deaf children. *OzCHI,* 135–142. https://doi.org/10.1145/3152771.3152786

Ladd, P. (2003). *Understanding deaf culture.* Multilingual Matters.

Lindtner, S., Anderson, K., & Dourish P. (2012). Cultural appropriation: Information technologies as sites of transnational imagination. In *Proceedings of the ACM Conference on Computer Supported Cooperative Work, CSCW* (pp. 77–86). https://doi.org/10.1145/2145204.2145220

Linxen, S., Cassau, V., & Sturm, C. (2021). Culture and HCI: A still slowly growing field of research. Findings from a systematic, comparative mapping review. In *ACM International Conference Proceeding Series.* https://doi.org/10.1145/3471391.3471421

Lu, F., Tian, F., Jiang, Y., Cao, X., Luo, W., Li, G., Zhang, X., Dai, G., & Wang, H. (2011). Shadow-Story: Creative and collaborative digital storytelling inspired by cultural heritage. In *Conference on Human Factors in Computing Systems-Proceedings* (pp. 1919–1928). https://doi.org/10.1145/1978942.1979221

Lu, Z., Annett, M., Fan, M., & Wigdor, D. (2019). "I feel it is my responsibility to stream" streaming and engaging with intangible cultural heritage through livestreaming. In *Conference on Human Factors in Computing Systems–Proceedings* (pp. 1–14). https://doi.org/10.1145/3290605.3300459

Lyu, Y., & Carroll, J. M. (2022). Cultural influences on chinese citizens ' adoption of digital contact tracing: A human infrastructure perspective. In *Conference on Human Factors in Computing Systems-Proceedings*.

Mankoff, J., Hayes, G. R., & Kasnitz, D. (2010). Disability studies as a source of critical inquiry for the field of assistive technology. In *ASSETS'10-Proceedings of the 12th International ACM SIGACCESS Conference on Computers and Accessibility* (pp. 3–10). https://doi.org/10.1145/1878803.1878807

Marsden, G., Maunder, A., & Parker, M. (2008). People are people, but technology is not technology. *Philosophical Transactions of the Royal Society A: Mathematical, Physical and Engineering Sciences, 366*(1881), 3795–3804. https://doi.org/10.1098/rsta.2008.0119

Mascarenhas, S., Degens, N., Paiva, A., Prada, R., Hofstede, G. J., Beulens, A., & Aylett, R. (2016). Modeling culture in intelligent virtual agents: From theory to implementation. *Autonomous Agents and Multi-Agent Systems, 30*(5), 931–962.

Mora-Cantallops, M., Muñoz, E., & Santamaría, R., & Sánchez-Alonso, S. (2021). Identifying communities and fan practices in online retrogaming forums. *Entertainment Computing, 38*, 100410. https://doi.org/10.1016/j.entcom.2021.100410

Nguyen, D. T., & Fussell S. R. (2012). How did you feel during our conversation? Retrospective analysis of intercultural and same-culture instant messaging conversations. In *Proceedings of the ACM Conference on Computer Supported Cooperative Work, CSCW* (pp. 117–126). https://doi.org/10.1145/2145204.2145225

Not, E., & Petrelli, D. (2019). Empowering cultural heritage professionals with tools for authoring and deploying personalised visitor experiences. *User Modeling and User-Adapted Interaction, 29*(1), 67–120. https://doi.org/10.1007/s11257-019-09224-9

Orji, R., & Mandryk, R. L. (2014). Developing culturally relevant design guidelines for encouraging healthy eating behavior. *International Journal of Human Computer Studies, 72*(2), 207–223. https://doi.org/10.1016/j.ijhcs.2013.08.012

Pereira, R., & Baranauskas, M. C. C. (2015). A value-oriented and culturally informed approach to the design of interactive systems. *International Journal of Human Computer Studies, 80*, 66–82. https://doi.org/10.1016/j.ijhcs.2015.04.001

Petrelli, D. (2019). Tangible interaction meets material culture: Reflections on the meSch project. *Interactions, 26*(5), 34–39. https://doi.org/10.1145/3349268

Prabhakar, A. S., Maris, E., & Medhi-Thies, I. (2021). Toward understanding the cultural influences on social media use of middle class mothers in india. In *Conference on Human Factors in Computing Systems-Proceedings*. https://doi.org/10.1145/3411763.3451779

Rau, P. L. P. (Ed.). (2018). *Cross-cultural design*. LNCS Springer.

Rau, P. L. P. (Ed.). (2020). *Cross-cultural design*. LNCS Springer.

Recabarren, M., & Nussbaum, M. (2010). Exploring the feasibility of web form adaptation to users' cultural dimension scores. *User Modeling and User-Adapted Interaction, 20*(1), 87–108. https://doi.org/10.1007/s11257-010-9071-7

Reinecke, K., Nguyen, M. K., Bernstein, A., Näf, M., & Gajos, K. Z. (2013). Doodle around the world: Online scheduling behavior reflects cultural differences in time perception and group decision-making. In *Proceedings of the ACM Conference on Computer Supported Cooperative Work, CSCW* (pp. 45–54). https://doi.org/10.1145/2441776.2441784

Ruiz-Calleja, A., Bote-Lorenzo, M. L., Asensio-Pérez, J. I., Villagrá-Sobrino, S. L., Alonso-Prieto, V., Gómez-Sánchez, E., García-Zarza, P., Serrano-Iglesias, S., & Vega-Gorgojo, G. (2023). Orchestrating ubiquitous learning situations about cultural heritage with casual learn mobile application. *International Journal of Human-Computer Studies, 170,* 102959. https://doi.org/10.1016/j.ijhcs.2022.102959

Russo, P., & Boor, S. (1993). How fluent is your interface? Designing for International Users. *Interchi,* 342–347.

Salgado, L., Faria, C., & Sieckenius de Souza, C. (2013). *A journey Through Cultures.* Springer Human-Computer Interaction Series.

Setlock, L. D. & Fussell S. R. (2011). Culture or fluency? unpacking interactions between culture and communication medium. In *Conference on Human Factors in Computing Systems-Proceedings* (pp. 1137–1140). https://doi.org/10.1145/1978942.1979112

Shao, H. (2021). Do i prefer it?: The role of cultural heuristics in chinese citizens' atitudes to COVID-19 rumors. In *Conference on Human Factors in Computing Systems–Proceedings.* https://doi.org/10.1145/3411763.3451647

Sheldon, P., Herzfeldt, E., & Rauschnabel, P. A. (2020). Culture and social media: the relationship between cultural values and hashtagging styles. *Behaviour and Information Technology, 39*(7), 758–770. https://doi.org/10.1080/0144929X.2019.1611923

Thayer, K., Guo, P. J., & Reinecke, K. (2018). The impact of culture on learner behavior in visual debuggers. In *Proceedings of IEEE symposium on visual languages and human-centric computing, VL/HCC, 2018-Octob* (pp. 115–124). https://doi.org/10.1109/VLHCC.2018.8506556

Triandis, H. (1995). *Individualism & Collectivism.* Routledge.

Vasalou, A., Joinson, A. N., & Courvoisier, D. (2010). Cultural differences, experience with social networks and the nature of "true commitment" in Facebook. *International Journal of Human Computer Studies, 68*(10), 719–728. https://doi.org/10.1016/j.ijhcs.2010.06.002

Vatrapu, R. K. (2010). Explaining culture: An outline of a theory of socio-technical interactions. In *Proceedings of the 3rd ACM International Conference on Intercultural Collaboration, ICIC '10* (pp. 111–120). https://doi.org/10.1145/1841853.1841871

Volkmar, G., Wenig, N., & Malaka, R. (2018). Memorial quest-a location-based serious game for cultural heritage preservation. In *CHI PLAY 2018 - Proceedings of the 2018 Annual Symposium on Computer-Human Interaction in Play Companion Extended Abstracts* (pp. 661–668). https://doi.org/10.1145/3270316.3271517

Wecker, A. J., Kuflik, T., & Stock, O. (2017). AMuse: Connecting indoor and outdoor cultural heritage experiences. In *International Conference on Intelligent User Interfaces, Proceedings IUI* (pp. 153–156). https://doi.org/10.1145/3030024.3040980

Zhou, S., Zhang, Z., & Bickmore, T. (2017). Adapting a persuasive conversational agent for the chinese culture. In *Proceedings-2017 International Conference on Culture and Computing, Culture and Computing 2017, 2017-Decem* (pp. 89–96). https://doi.org/10.1109/Culture.and.Computing.2017.42

Some Open Issues

6

This chapter discusses some open issues in HCI research on (and with) culture. These issues are classified into three (non-mutually exclusive) categories: theoretical, practical, and controversial. Regarding theoretical issues, this chapter discusses two opposing perspectives of the concept of culture adopted in HCI, i.e., culture as a definitive or a sensitizing concept, and the need to reflect further on the theoretical/conceptual perspective. With respect to practical issues, this chapter suggests that further work is warranted to better understand cultural differences and similarities. There is also a need to work with other user groups than students and go beyond equating nationalities with culture to address cultural diversity. Bringing together taxonomic and contingent perspectives might be a way forward towards addressing this goal. In controversial issues, this chapter raises the question of whether problems or issues brought about by taking into account the cultural dimension are or can be always addressed with a technological solution. This chapter also engages in a discussion on the extent to which we should accept and/or deal with the fact that culture is a stereotyping mechanism. Finally, this chapter also discusses the complexity of designing technologies for culturally diverse users.

6.1 Theoretical Issues

6.1.1 The Concept of Culture: A Definitive or a Sensitizing One?[1]

HCI research on culture sees the concept of culture in two (very) different views. The taxonomic perspective sees culture as a definitive concept. The Cultural Dimensions of Hofstede (2010), the cross-cultural theory of communication of Hall (1973, 1976, 1982),

---

[1] This open issue is inspired by Hertzum's commentary on usability as a sensitizing concept (Hertzum, 2018). This section adapts his commentary to the context of culture within HCI research.

© The Author(s), under exclusive license to Springer Nature Switzerland AG 2023 71
S. Sayago, *Cultures in Human-Computer Interaction*, Synthesis Lectures
on Human-Centered Informatics, https://doi.org/10.1007/978-3-031-30243-5_6

and the cultural cognitive systems of thought of Nisbett (2003) are the main conceptual perspectives of much of this research. The contingent perspective, however, views culture as a sensitizing concept, looking at culture as a dynamic, constructed, and interactional phenomenon. Both conceptualizations have advantages and disadvantages.

A definitive concept of culture employs a clear definition and identification of its main attributes. This conceptualization enables researchers to speak a common language (i.e., we are referring to the same concept) and to study culture in a consistent way. However, a definitive concept does not allow researchers to be open and flexible. We might run the risk of overlooking important factors not included in the definition, studying something which is not especially relevant in the case required, or focusing more on conceptual precision than on practical relevance.

A sensitizing concept of culture addresses the risks of the definitive conceptualization of the term by recognizing and dealing with diversity. Culture has different definitions because it means different things depending on the case studied. The risks associated with a sensitizing concept of culture include ambiguity, that comparisons are hard, and that conceptual and theoretical development is difficult or too individualized to become useful in a general case.

This book leans towards understanding culture as a sensitizing concept. To study technical issues, well-defined concepts are needed. For instance, in Computer Science, it is important to define unambiguously and in detail concepts like units of storage (e.g., 1 Kb = 1024 bytes) or what a programming instruction does exactly (e.g., SELECT * FROM table_name WHERE condition) to design, develop, and evaluate algorithms, hardware and software applications. If the objective is the empirical study of culture within HCI, definitive concepts of this term might be not so useful or needed, or might even become counterproductive, because of the constructed and interactional aspects of both culture and technology. In any case, rather than thinking of one type of concept of culture as being superior to the other, HCI researchers could reflect further on the pros and cons of each conceptualization.

6.1.2 Reflecting Further on Theory

The *Cultural Dimensions* (Hofstede et al., 2010) is a valuable piece of work on culture for both scholars and practitioners (Jones, 2007). Chapter 5 shows this model of culture is the most predominant in HCI. The empirical basis is (very) strong. Yet, the overwhelming influence of this work[2] has perhaps made it too easy for other researchers (HCI included)

---

[2] As opposed to, for instance, the theory of types of cultural value orientations put forward by Schwartz (2004), which has received less research attention than Hofstede's Cultural Dimensions. Yet, it draws on data gathered in 1988–2000 with 80 samples of school teachers (k-12) from 58 national groups and 115 samples of college students from 64 national groups, together constituting 67 nations and 70 different.

to use the Cultural Dimensions in an uncritical manner (Javidan et al., 2006). Consequently, the level of controversy surrounding this work is quite high (Jones, 2007). Some issues are outlined and contextualized next.[3]

The Cultural Dimensions of Hofstede were developed within the context of Business and Management, wherein seeing culture as solving problems makes sense. How do we make a deal/do business with a person from a different culture from our own? Is there just one way of managing a company? How do we train our employees to work abroad (in a country with a culture different from ours)? As stated in (Jones, 2007), during the time of the delivery of the Cultural Dimensions, there was very little work on culture in Business and Management, and at this time many businesses were just entering the international arena and were experiencing difficulties. The Cultural Dimensions provided them with (arguably) credible advice. Cross-cultural training based on an empirically derived conceptual scheme turned out to be more effective than training based on ideas and beliefs (Triandis, 2007).

As discussed in (Baskerville, 2002), Cultural Dimensions atomizes culture into bits and pieces, considers each element to be of equal significance, and poses the risk of stereotyping. The approach assumes that the presence of that cultural element in one society is equivalent to the presence of that cultural element in another. The approach also assumes national uniformity, that questionnaire responses from some employees in a single organization are also the national average tendency (McSweeney, 2002), and that culture is static. The importance of 'being within' is also missing in Hofstede's work. Cultural dimensions cannot tell us how cultural values are put into practice (Kamppuri, 2012). This work also seems to perpetuate a prevalent sense of confidence that all dimensions of national culture have been discovered (Javidan et al., 2006). For example, the vertical axis (the North vs the South) does not seem to have been considered in the Cultural Dimensions, which focus on the horizontal one (the East vs. the West). How would the dimensions change if we placed nations on a vertical axis? (Kitayama & Cohen, 2007).

Culture as solving problems can also be appreciated in the cross-cultural theory of communication of Hall, as some of the aspects discussed (e.g., monochromic or polychromic time, high- and low-context culture) can help us to avoid and deal with cultural misunderstandings, which are very important nowadays. Hall focuses on drawing our attention to what we communicate without the use of words (e.g., space and time talks) and to the need of understanding our (as well as others') culture/s better. This focus on seeing culture as communication can also be accounted for by the context in which it is grounded: cross-cultural training. In a similar way as Hofstede, Hall seems to propose principles which compose a kind of how-to guide for identifying and solving intercultural communication issues. However, as opposed to Hofstede in *Cultural Dimensions* (2010), Hall considers that no culture exists exclusively on one end of a scale. This diversity is also echoed by Triandis in *Individualism and Collectivism*, wherein it is

---

[3] Whether these issues are attributable to the model *per se* and / or the way in which it has been adopted and used is out of the scope of this synthesis.

argued that none of us is a 'pure' individualist or collectivist, as this decision is mostly situation-specific, and depends on a number of factors, such as tightness/looseness, and complexity/simplicity.

As outlined in Chap. 5, HCI research on cultures confirms these conceptual perspectives. HCI does not seem to add anything new or different from a theoretical perspective. The Cultural Dimensions of Hofstede (Hofstede et al., 2010) and the Cross-Cultural Theory of Communication of Hall (1973, 1976, 1982) date back to the 20th century. The cultural cognitive systems of thought of Nisbett (2003) is slightly more recent, early 21st century. None of them has been developed for designing and evaluating digital technologies for human use. It might well be that these works–despite well-recognized limitations and criticism–stand the test of time and/or are valid within HCI. On the other hand, it might be that HCI research on culture still has to reflect further on its theoretical component.

6.2 Practical Issues

6.2.1 Understanding Further the Differences (and Similarities)

Virtually all studies have found differences. What leads to these differences is less clear. To say that 'communication differences between American and Chinese university students can be attributed to differences in communication styles (low vs. high context), or that 'websites aimed at high uncertainty-avoidance cultures should give prominence to leaders, focus on expertise and authority, and provide structured access to information' is useful but not enough to either understand our users better or design technologies that cater for their needs and add value to their lives. Most of the studies account for the differences in the results by drawing on the conceptual perspectives outlined in Chap. 4. However, there are different ways in which one can be individualist or collectivist. People who belong to individualist/collectivist or high/low uncertainty-avoidance cultures can become individualist (or collectivist), or avoid uncertainty, in a myriad of ways. These cultural traits are also elicited by particular situations, i.e., people are not always individualist or collectivist, and no culture is monolithic. The dimensional approach focuses on quantifying differences and classifying countries. It says little about the process by which these differences constitute each other (Rose & Yam, 2007). As stated in Chap. 2, culture always emerges in interaction. Cultural differences are not located solely within individuals but within practices, situations, collective representations, and so on. Thus, there might be other underlying reasons that provide an explanation for cultural differences than those on which the 'light is already shinning and the territory is well illuminated'(Heine, 2016).

In addition to delving into what makes us different, for HCI research on culture to grow and mature, further attention to similarities might be needed. Each person may be described in three ways: the universal characteristics of the species, the sets of characteristics that define that person as a member of a series of groups, and that person's idiosyncratic characteristics. When we talk of traits which are neither universal nor idiosyncratic we use the term culture (Wallerstein, 1990). What exactly makes us similar (or we have in common) as far as interacting with digital technologies is concerned? And what is the impact of these commonalities in HCI research? For example, in people who have a bi-cultural background, differences on field dependence (i.e., processing information more holistically or analytically (Nisbett, 2003)) between cultures might be reduced, making people with different national cultures equally (i.e., non-significant differences) susceptible to distraction by a secondary-task when filling out online psychometric questionnaires (van Schaik et al., 2015). Also, some users might have more exposure to English language than English speakers do to non-English websites, and this difference in exposure might account for similarities in cross-cultural performance and satisfaction with websites (Alexander et al., 2021).

6.2.2 Beyond Students and Nationalities: Towards More Diversity

The profile of research outlined in Chap. 5 shows that HCI research on culture is mostly quantitative, cross-cultural/comparative, and conducted in few nations. The participants tend to be university students and much research relates nations or countries with cultural groups. Students might be easy to reach for researchers and this might explain why the participants of studies tend to be students. Equating nation with culture might also simplify things considerably, and it is something done in other areas, such as marketing, wherein we can talk about national markets, e.g., the Chinese market (Singh & Pereira, 2005). Yet, due to the fact that diversity is an important issue in the discourses on cultures, there is a need to work with a more diverse group of participants and consider other visions of culture than national groups (e.g., age groups, subcultural groups, and minorities). How well do the results of studies conducted with students apply to other user groups, such as children and older adults? Children can be regarded as cultural learners whereas older people are cultural masters. There are also subcultural groups and minorities. We need to work with more diverse and hybrid user groups. Further studies examining cultures from 'within', adopting a qualitative research approach, can contribute to the study of cultural issues in HCI more deeply.

6.2.3 Bringing Together Taxonomic and Contingent Perspectives: A Way Forward?

Bringing together the taxonomic and contingent perspectives might be a way forward to deal with cultural diversity in HCI, as doing so combines benefits from the richness and specificity gained from qualitative studies examining a cultural group from 'within', with the pragmatism of conceptualizing culture as a set of dimensions offered by cultural models. (Halabi and Zimmermann, 2019) puts forwards bridging together both perspectives by developing categories based on a specific understanding of the physical world (in the spirit of the taxonomic perspective), while at the same time taking into account the subjectivity, the locality, and the dynamics of the data collected (in the spirit of the contingent perspective). By doing so, it might be possible to take advantage of both taxonomic and contingent perspectives. The taxonomic perspective is useful to discuss multiple instances of expressions and interactions, for example, a similar pattern of behavior among a group of people. It allows us to classify these expressions and interactions in different categories. The contingent approach, on the other hand, is useful to discuss the dynamic changes in expressions and interactions. (Alshehri et al., 2021) propose Qualitative Secondary Analysis (QSA) as an alternative approach to transcend the limitations of both time and resource intensive ethnographies (contingent perspective) and universal models of culture (taxonomic perspective). The proposed approach combines benefits from the richness and specificity gained from ethnography with the pragmatism of conceptualizing culture as a set of dimensions offered by cultural models. QSA is an empirical exercise conducted by the re-use of existing data–produced in previous (primary) studies. (Alshehri et al., 2021) argues that qualitative datasets have much to offer beyond answering the primary research questions, to explore aspects that may have never been analyzed. By conducting QSA in two case studies with Saudi transnationals, (Alshehri et al., 2021) show that QSA can provide a deeper meaning on how cultural dimensions such as collectivism is constructed uniquely in a given culture. QSA is not intended to replace universal cultural models, rather it is meant to provide a close lens to examine the specificity of each culture.

6.3   Controversial Issues

6.3.1   Solutions to Cultural Issues: Always Technological?

After reading this synthesis, one might get the impression that culture brings about a number of problems or issues. As such, solutions have been developed. For example, researchers are faced with challenges, such as failing to understand the cultural nuances of a community (e.g., hierarchies and local norms) when working with users with a different cultural background than theirs, and a non-academic partnership between researchers and users is needed (Sabie et al., 2022). There is a need to localize and contextualize usability because usability does not mean the same for all of us. Also, usability evaluations are not performed in the same way 'here and there'. In order to operationalize culture in design, HCI designers need support. Tools, toolkits, design guidelines, and approaches have been proposed. There is a cultural question in Participatory Design (PD) too. The ways of doing and appropriating PD differs across culture. It is also unreasonable to design one common website (or application) for everyone, and yet expect to attract an international audience. A more desirable solution seems to be to design many different versions of a user interface in order to cater for users' cultural preferences. If this discourse on "problems or issues and technological solutions" is read within the context of technological solutionism (Morozov, 2013), it brings up the question of whether the solution to cultural issues in HCI is always technological. For example, people with very different communication and working styles–because of different cultural background profiles–are able to adapt and overcome communication and coordination issues without a technological aid, or using the same technologies. When does a cultural issue not need a technological solution in HCI, and why? This question might be provocative because HCI is at its core about computers and human interaction. Why focus on non-technological solutions? Just as examining technology non-use reveals alternative meanings of technologies and problematic consequences of these technologies (Knowles & Hanson, 2018), which are important aspects for HCI, exploring when a cultural issue does not need a technological solution enables reflection on the reasons for which a technological intervention is not needed, and this enriches our understanding of computers and human interaction.

6.3.2 Culture as a Stereotyping Mechanism

Conducting HCI research on (and with) cultures is not free from potential dangers. (Buchtel, 2014) poses a number of questions within the context of teaching about cultural differences in Cultural Psychology that are relevant for the HCI community too. Might HCI research on culture encourage stereotyping when thinking about cultures or individuals? Can it be misinterpreted as legitimizing cultural stereotypes? Could it even increase prejudice, at least for some people? Culture is the vehicle through which stereotypic knowledge is transferred within a society and across generations (Williams & Spencer-Rodgers, 2010). Human functioning requires cognitive categorization because of information overload (Cuddy & Fiske, 2002). We cannot avoid this cognitive and social process because it is how the mind and social groups organize and categorize their perceptions (Williams & Spencer-Rodgers, 2010). Once formed, the categories become the basis for individual and social prejudgment and stereotypes (Thornton, 2002). Culture is or can become a stereotyping mechanism. Perhaps, this is something that needs to be accepted. Yet, this might not be taken as an excuse to avoid inaccurate stereotyping or misunderstandings. For example, the very idea of a stereotype, that an individual is perceived through the lens of his or her social group membership, may not be perceived negatively by members of cultural groups who strongly value their group memberships and more readily adhere to their groups' norms (Williams & Spencer-Rodgers, 2010). HCI research on culture has great potential to help–and harm. Awareness of positive and negative effects can help us explicitly address these issues when doing HCI research in which culture is a primary topic of interest.

6.3.3 Cultural Design

How can a technology be designed as both usable and meaningful to culturally diverse users? (Sun, 2012) Chap. 3 shows that culture matters in technology design. Yet, the answer to the preceding question is not clear and far from being straightforward. We could consider an approach that emphasizes cultural uniqueness and authenticity, e.g., ethnography. Yet, this may encourage us to design N interfaces, one for each cultural group (or subgroup) (Aykin, 2005). We could also think about internationalization and localization, which are widely used (Salgado, Faria & Sieckenius de Souza, 2013; Singh, 2005). However, cross-cultural technology design is not just about translating a dialogue box or localizing an icon, it is a process of transferring meaning (Sun, 2012). In this sense, we could turn our attention to culturalization, which is gaining traction in video games, and is a deeper level of game localization, which includes changing game context and contents that are suitable for local customs, traditions, religion, beliefs, rules, and laws (Pyae, 2018; Honeywood & Fung, 2012). Other design approaches outlined in Chap. 3, such as cultural (automatic) adaptation, tools and toolkits and like Cultural Viewpoint Metaphors (CVM) (Salgado, Faria & Sieckenius de Souza, 2013) and Cultura (Hao, Van Boeijen & Stappers, 2017), can also be used. Currently, deep learning techniques are also being explored to support designers in cross-cultural design (Zhou et al., 2022). Which one should be adopted (and why)? Perhaps, there is not any single approach to designing culturally-meaningful user interfaces, because it depends on the objectives we have and each empirical instance. Yet, it seems clear that simply translating the text of a user interface, or changing their colors, is not enough for transferring meaning. It is also clear that focusing only on a cultural group is not useful enough either (especially in terms of cross-cultural products). We need to design for both the visible and invisible parts of the iceberg metaphor (see Chap. 2). There is widespread agreement on this assertion, but beyond that, the question raised at the opening of this section does not have a definitive answer–perhaps, if we think of culture as a sensitizing concept, the answer to the question above is or should be a sensitizing one too.

References

Alexander, R., Thompson, N., McGill, T., & Murray, D. (2021). The influence of user culture on website usability. *International Journal of Human-Computer Studies, 154*, 102688. https://doi.org/10.1016/j.ijhcs.2021.102688

Alshehri, T., Abokhodair, N., Kirkham, R., & Olivier, P. (2021). Qualitative secondary analysis as an alternative approach for cross-cultural design a case study with saudi transnationals. *Conference on Human Factors in Computing Systems-Proceedings.* https://doi.org/10.1145/3411764.3445108

Aykin, N. (Ed.). (2005). *Usability and internationalization of information technology.* Lawrence Erlbaum Associates Publishers. https://doi.org/10.1109/memb.2006.1578651

Baskerville, R. F. (2002). Hofstede never studied culture. *Accounting, Organizations and Society, 28*(1), 1–14. https://doi.org/10.1016/S0361-3682(01)00048-4

Buchtel, E. E. (2014). Cultural sensitivity or cultural stereotyping? Positive and negative effects of a cultural psychology class. *International Journal of Intercultural Relations, 39*, 40–52. https://doi.org/10.1016/j.ijintrel.2013.09.003

Cuddy, A., & Fiske, S. (2002). Doddering but dear: Process, content, and function in stereotyping of older persons. In T.D. Nelson, T.D (Ed.), *Ageism. Stereotyping and Prejudice against Older Persons* (pp. 3-27). MIT Press

Halabi, A., & Zimmermann, B. (2019). Waves and forms: constructing the cultural in design. *AI and Society, 34*(3), 403–417. https://doi.org/10.1007/s00146-017-0713-8

Hall, E. (1973). *The silent language.* Anchor Books.

Hall, E. (1976). *Beyond culture.* Anchor Books.

Hall, E. (1982). *The hidden dimension.* Anchor Books.

Hao, C., Van Boeijen, A., & Stappers, P. J. (2017). Cultura: A communication toolkit for designers to gain empathic insights across cultural boundaries. In *Proceedings of the IASDR Conference RE: Research* (pp. 497–510). https://doi.org/10.7945/C2SD5J

Heine, S. (2016). *Cultural psychology* (3rd ed.). W. W. Norton & Company.

Hertzum, M. (2018). Commentary: Usability—a sensitizing concept. *Human-Computer Interaction, 33*(2), 178–181. https://doi.org/10.1080/07370024.2017.1302800

Hofstede, G., Hofstede, G. J., & Minkov, M. (2010). *Cultures and organizations.* McGraw Hill.

Honeywood, R., & Fung, J. (2012). Best practices for game localization. In IGDA Game Localization SIG. http://www.igda.org/members/group_content_view.asp?group=121045&id=526230

Javidan, M., House, R. J., Dorfman, P. W., Hanges, P. J., & De Luque, M. S. (2006). Conceptualizing and measuring cultures and their consequences: A comparative review of GLOBE's and Hofstede's approaches. *Journal of International Business Studies, 37*(6), 897–914. https://doi.org/10.1057/palgrave.jibs.8400234

Jones, M. L. (2007). Hofstede-culturally questionable?. Oxford Business & Economics Conference.

Kamppuri, M. (2012). Because deep down, we are not the same: Values in cross-cultural design. *Interactions, 19*(2), 65–68. https://doi.org/10.1145/2090150.2090166

Kitayama, S., & Cohen, D. (Eds.). (2007). *Handbook of cultural psychology.* The Guildford Press.

Knowles, B., & Hanson, V. (2018). The wisdom of older technology (non)users. *Communications of the ACM, 61*(3) (March 2018), 72–77. https://doi.org/10.1145/3179995

McSweeney, B. (2002). *Hofstede's model of national cultural differences and their consequences: A triumph of faith—a failure of analysis.* Human Relations.

Morozov, E. (2013). *To save everything, click here.* Public Affairs.

Nisbett, R. (2003). The geography of thought. *The Free Press.* https://doi.org/10.1111/j.0033-0124.1973.00331.x

Pyae, A. (2018). Understanding the role of culture and cultural attributes in digital game localization. *Entertainment Computing, 26*, 105–116. https://doi.org/10.1016/j.entcom.2018.02.004

Rose, H., & Yam, M. (2007). Sociocultural Psychology: The Dynamic Interpendence among self systems and social systems. In S. Kitayama & D. Cohen (Eds.), *2007* (pp. 3–40). The Guildford Press.

Sabie, D., Ekmekcioglu, C., & Ahmed, S. I. (2022). A Decade of international migration research in HCI: overview, challenges, ethics, impact, and future directions. *ACM Transactions on Computer-Human Interaction, 29*(4), 1–35. https://doi.org/10.1145/3490555

Salgado, L., Faria, C., & Sieckenius de Souza, C. (2013). *A journey through cultures.* Springer Human-Computer Interaction Series.

Schwartz, S. H. (2004). Mapping and interpreting cultural differences around the world. In H. Vinken, J. Soeters, & P Ester (Eds.), *Comparing cultures, Dimensions of culture in a comparative perspective* (pp. 43-73). Leiden, The Netherlands: Brill.

Singh, N., & Pereira, A. (2005). *The Culturally customized web site.* Elsevier.

Sun, H. (2012). *Cross-cultural technology design. Crafting culture-sensitive technology for local users.* Oxford Series in Human-Technology Interaction.

Thornton, J. E. (2002). Myths of aging or ageist stereotypes. *Educational Gerontology, 28,* 301–312.

Triandis, H. (2007). Culture and Psychology: A history of the study of their relationship. In S. Kitayama & D. Cohen (Eds.), *2007* (pp. 59–77). The Guildford Press.

van Schaik, P., Luan Wong, S., & Teo, T. (2015). Questionnaire layout and national culture in online psychometrics. *International Journal of Human-Computer Studies, 73,* 52–65. https://doi.org/10.1016/j.ijhcs.2014.08.005

Wallerstein, I. (1990). Culture as the ideological battleground of the modern world-system. *Theory, Culture & Society, 7*(2–3), 31–55. https://doi.org/10.1177/026327690007002003

Williams, M. J., & Spencer-Rodgers, J. (2010). Culture and stereotyping processes: Integration and new directions: Culture and stereotyping. *Social and Personality Psychology Compass, 4*(8), 591–604. https://doi.org/10.1111/j.1751-9004.2010.00288.x

Zhou, L., Sun, X., Mu, G., Wu, J., Zhou, J., Wu, Q., Zhang, Y., Xi, Y., Gunes, N. D., & Song, S. (2022). A tool to facilitate the cross-cultural design process using deep learning. *IEEE Transactions on Human-Machine Systems, 52*(3), 445–457. https://doi.org/10.1109/THMS.2021.3126699

Conclusion

7

This book aims to contribute to HCI with a structured overview of a great deal of HCI research in which culture is a significant theme. Much of this research is scattered and focuses on particular topics of interest. This situation hinders the developing of an overall view of HCI research on (and with) culture. This bird's-eye view aids in providing an integrated answer to important research questions for a field to develop; in this case, why culture is important in HCI, what has been done thus far, and where we can go from there. Working towards this goal, the research literature this book draws on is interdisciplinary, including over 40 books and 400 conference and journal papers published in several academic databases. Although this body of knowledge is not comprehensive enough, it allows the synthesis to attain its objective, as outlined in this chapter.

Culture is an important and complex concept. Culture is an important concept because it impacts every aspect of our lives. It is in this relevance, which can be seen from the many and varied definitions of culture, where the complexity of the concept of culture lies, as its meaning is contextual and fragmentary. Culture can be seen from two opposing perspectives: culture as a sensitizing or a definitive concept. Yet, there is widespread agreement in the literature that culture is learned, constructed, shared, and dynamic. There is also increasing awareness of the fact that culture is not synonymous with nationality, what we get when we study arts, visit a museum, listen to classical music, or just what we see. The relevance and complexity of culture are unlikely to become weaker, less intricate, or disappear in the near future. Technologies, wars, environmental and health crises come and go and make culture change, but cultures persist. We cannot do without culture.

Despite the relevance of culture to study the rich and complex nature of our lives, there are reasons to be skeptical about the importance of culture in HCI. The moderate interest of the HCI community in culture might be accounted for the fact that the concept defies a single and unique definition, and it is constantly changing. Further reasons include globalization and the perceived added value of culture to HCI when the field already

© The Author(s), under exclusive license to Springer Nature Switzerland AG 2023
S. Sayago, *Cultures in Human-Computer Interaction*, Synthesis Lectures on Human-Centered Informatics, https://doi.org/10.1007/978-3-031-30243-5_7

addresses values, identity, emotions, and experiences. However, there are also reasons for claiming that culture matters (a lot) in HCI, including a diversely growing user base– and consequently a lack of global users, the need for providing designers with enough support to design across cultures and to adapt usability evaluation methods to several cultural realities, and the inseparable relationship between culture and technology. Culture impacts the three *sine qua non* components of this field: Humans–Computers–Interaction.

The Cultural Dimensions of Hofstede, the cross-cultural theory of communication of Hall, and the cultural cognitive systems of thought of Nisbett, predominate in much HCI research. None of these conceptual frameworks has been developed for designing and evaluating digital technologies for human use. HCI has a long tradition of incorporating knowledge from other disciplines and this carries over to HCI research on culture too. Much of this research draws upon these conceptual perspectives to account for the results and/or discuss the design of empirical studies. Beyond that, much of this research does not add anything new or different from a theoretical perspective. To provide a richer conceptual perspective, this book (in Chap. 3) has also outlined other views of culture which, despite not playing a central role in HCI, deepen and extend the main ones.

The book brings together individual analyses of different periods of research and sources to provide an integrated profile of investigation. Much research is quantitative, conducted with university students in few nations, and cross-cultural/comparative. Two main approaches to address culture in HCI are adopted. The taxonomic perspective renders human ways of living and thinking into a finite set of elements sorted into categories. The contingent perspective looks at culture as a dynamic, constructed, and interactional phenomenon. Culture has largely been used in HCI to understand and account for the ways in which people interact with digital technologies, summarize the ways in which groups of people distinguish themselves from other groups, deal with and reduce the complexity of people's interactions with technologies, and find inspiration and foster reflection.

HCI research on (and with) culture is developing. Beyond that, it is difficult to draw other firm conclusions that much more research is warranted to better understand the culture within HCI. For HCI to grow and mature, this book has put forward a number of open issues, from the need to reflect on the conceptualization of culture and theoretical aspects, engaging further with diversity and to go beyond students, nationalities, and differences, to better understanding the benefits and perils of doing research on culture in HCI, and how to effectively design culturally meaningful technologies for a growing and heterogeneous user base. This synthesis (and author) look forward to the dialogue that addressing these and other related issues will spur about how to move forward with the concept of culture in HCI.

Bibliography

References that are not cited in the chapters are listed here:

Abdelnour-Nocera, J., & Rangaswamy, N. (2018). Reflecting on the design-culture connection in HCI and HCI4D. *Interactions, 25*(5), 8–9. https://doi.org/10.1145/3264381

Abdelnour-Nocera, J., Smith, A., Moore, J., Oyugi, C., Camara, S., Ressin, M., Shresta, S., & Wiles, A. (2011). The centre for internationalization and usability: Enabling culture-centred design for all. In P. Campos, N. Graham, J. Jorge, N. Nunes, P. Palanque, & M. Winckler (Eds.), *Human-computer interaction – INTERACT 2011* (Vol. 6949, pp. 683–684). Springer. https://doi.org/10.1007/978-3-642-23768-3_114

Abeele, M. V., & Roe, K. (2011). New life, old friends: A cross-cultural comparison of the use of communication technologies in the social life of college freshmen. *Young, 19*(2), 219–240. https://doi.org/10.1177/110330881001900205

Agreste, S., De Meo, P., Ferrara, E., Piccolo, S., & Provetti, A. (2015). Analysis of a heterogeneous social network of humans and cultural objects. *IEEE Transactions on Systems, Man, and Cybernetics: Systems, 45*(4), 559–570. https://doi.org/10.1109/TSMC.2014.2378215

Ahtinen, A., Ramiah, S., Blom, J., & Isomursu, M. (n.d.-a). *Design of mobile wellness applications: Identifying cross-cultural factors.*

Ai He, H., Memarovic, N., Sabiescu, A., & De Moor, A. (2015). CulTech2015: Cultural diversity and technology design. In C&T, 27–30 June (pp. 153–156). https://doi.org/10.1145/2768545.2768561

Airoldi, (2022). *Machine habitus. Toward a sociology of algorithms.* Polity

Alexander, R., Murray, D., & Thompson, N. (2017a). Cross-cultural web design guidelines. In *Proceedings of the 14th International Web for All Conference* (pp. 1–4). https://doi.org/10.1145/3058555.3058574

Alexander, R., Thompson, N., & Murray, D. (2017b). Towards cultural translation of websites: A large-scale study of Australian, Chinese, and Saudi Arabian design preferences. *Behaviour & Information Technology, 36*(4), 351–363. https://doi.org/10.1080/0144929X.2016.1234646

Alexander, R., Thompson, N., McGill, T., & Murray, D. (2021). The influence of user culture on website usability. *International Journal of Human-Computer Studies, 154*, 102688. https://doi.org/10.1016/j.ijhcs.2021.102688

Aljaroodi, H. M., Adam, M. T. P., Teubner, T., & Chiong, R. (2022). Understanding the Importance of cultural appropriateness for user interface design: An avatar study. *ACM Transactions on Computer-Human Interaction, 29*(6), 1–27. https://doi.org/10.1145/3517138

Almohamed, A., & Vyas, D. (2019). Rebuilding social capital in refugees and asylum seekers. *ACM Transactions on Computer-Human Interaction, 26*(6), 1–30. https://doi.org/10.1145/3364996

S. Sayago, *Cultures in Human-Computer Interaction*, Synthesis Lectures on Human-Centered Informatics, https://doi.org/10.1007/978-3-031-30243-5

AlMuhanna, N., Hall, W., & Millard, D. E. (2022). Fear of the dark: A cross-cultural study into how perceptions of antisocial behaviour impact the acceptance and use of Twitter. *Behaviour & Information Technology, 1–14.* https://doi.org/10.1080/0144929X.2022.2064766

Alshehri, T., Abokhodair, N., Kirkham, R., & Olivier, P. (2021). Qualitative secondary analysis as an alternative approach for cross-cultural design: A case study with Saudi transnationals. In *Proceedings of the 2021 CHI Conference on Human Factors in Computing Systems* (pp. 1–15). https://doi.org/10.1145/3411764.3445108

Alshehri, T., Kirkham, R., & Olivier, P. (2020). Scenario co-creation cards: A culturally sensitive tool for eliciting values. In *Proceedings of the 2020 CHI Conference on Human Factors in Computing Systems* (pp. 1–14). https://doi.org/10.1145/3313831.3376608

AlSulaiman, S., & Horn, M. S. (2015). Peter the fashionista?: Computer programming games and gender oriented cultural forms. In *Proceedings of the 2015 Annual Symposium on Computer-Human Interaction in Play* (pp. 185–195). https://doi.org/10.1145/2793107.2793127

Altarriba Bertran, F., Duval, J., Isbister, K., Wilde, D., Márquez Segura, E., Garcia Pañella, O., & Badal León, L. (2019). Chasing play potentials in food culture to inspire technology design. In *Extended Abstracts of the Annual Symposium on Computer-Human Interaction in Play Companion Extended Abstracts* (pp. 829–834). https://doi.org/10.1145/3341215.3349586

Amant, K. S. (2015). *Introduction to the special issue cultural considerations for communication design: Integrating ideas of culture, communication, and context into user experience design.*

Ankenbauer, S. A., & Lu, A. J. (2022). Making space for cultural infrastructure: The breakdown and maintenance work of independent movie theaters during crisis. In *CHI Conference on Human Factors in Computing Systems* (pp. 1–13). https://doi.org/10.1145/3491102.3501840

Annamoradnejad, I., Fazli, M., Habibi, J., & Tavakoli, S. (2019). Cross-cultural studies using social networks data. *IEEE Transactions on Computational Social Systems, 6*(4), 627–636. https://doi.org/10.1109/TCSS.2019.2919666

Appadurai, A. (2009). *The social life of things: Commodities in cultural perspective* (7th ed.). Cambridge University Press.

Aragon, C. R., & Poon, S. (2011). No sense of distance: Improving cross-cultural communication with context-linked software tools. In *Proceedings of the 2011 IConference* (pp. 159–165). https://doi.org/10.1145/1940761.1940783

Arawjo, I. (2020). To write code: The cultural fabrication of programming notation and practice. In *Proceedings of the 2020 CHI Conference on Human Factors in Computing Systems* (pp. 1–15). https://doi.org/10.1145/3313831.3376731

Ardissono, L., Kuflik, T., & Petrelli, D. (2012). Personalization in cultural heritage: The road travelled and the one ahead. *User Modeling and User-Adapted Interaction, 22*(1–2), 73–99. https://doi.org/10.1007/s11257-011-9104-x

Arvila, N., Fischer, A., Keskinen, P., & Nieminen, M. (2018). Mobile weather services for Maasai farmers: Socio-cultural factors influencing the adoption of technology. In *Proceedings of the Second African Conference for Human Computer Interaction: Thriving Communities* (pp. 1–11). https://doi.org/10.1145/3283458.3283466

Asokan, A. (2008). Culture calling: Where is CHI? In *CHI'08 Extended Abstracts on Human Factors in Computing Systems* (pp. 2383–2386). https://doi.org/10.1145/1358628.1358690

Avram, G., & Maye, L. (2016). Co-designing encounters with digital cultural heritage. In *Proceedings of the 2016 ACM Conference Companion Publication on Designing Interactive Systems* (pp. 17–20). https://doi.org/10.1145/2908805.2908810

Aykin, N. (Ed.). (2005). *Usability and internationalization of information technology.* Lawrence Erlbaum Associates Publishers. https://doi.org/10.1109/memb.2006.1578651

Baeza-Yates. R. (2018, June). Bias on the web. *Communications of the ACM, 61*(6), 54–61. https://doi.org/10.1145/3209581

Baldwin, J. R. (Ed.). (2006). *Redefining culture: Perspectives across the disciplines.* Lawrence Erlbaum Associates.

Baldwin, J. R., Faulkner, S. L., Hecht, M. L., & Lindsley, S. L. (Eds.). (2006). *Redefining culture: Perspectives across the disciplines.* Lawrence Erlbaum Associates Publishers. https://doi.org/10.4324/9781410617002

Bannon, L. (2011). Reimagining HCI: Toward a more human-centered perspective. *Interactions, 29,* 50–57.

Bannon, L. J. (1991). From human factors to human actors the role of psychology and human-computer interaction studies in systems design. In *Design at work: Cooperative design of computer systems* (pp. 25–44).

Baranoski, J., Kleinman, E., Wang, Z., Tucker, M., Ahnert, J., Schell, J., Doshi, R., Rank, S., & Zhu, J. (2015). Matsya: A cultural game of flow and balance. In *Proceedings of the 2015 Annual Symposium on Computer-Human Interaction in Play* (pp. 751–754). https://doi.org/10.1145/2793107.2810272

Bardzell, S. (2018). Utopias of participation: Feminism, design, and the futures. *ACM Transactions on Computer-Human Interaction, 25*(1), 1–24. https://doi.org/10.1145/3127359

Bardzell, S., Bardzell, J., & Ng, S. (2017). Supporting cultures of making: Technology, policy, visions, and myths. In *Proceedings of the 2017 CHI Conference on Human Factors in Computing Systems* (pp. 6523–6535). https://doi.org/10.1145/3025453.3025975

Barker, C., & Jane, E. (2016). *Cultural studies: Theory and practice.* SAGE.

Barrett, L. (2017). *How emotions are made: The secret life of the brain.* Houghton Mifflin Harcourt

Barricelli, B. R., Fischer, G., Fogli, D., Mørch, A., Piccinno, A., & Valtolina, S. (2016). Cultures of participation in the digital age: From "Have to" to "Want to" participate. In *Proceedings of the 9th Nordic Conference on Human-Computer Interaction* (pp. 1–3). https://doi.org/10.1145/2971485.2987668

Bartneck, C., Nomura, T., Kanda, T., Suzuki, T., & Kato, K. (2005). *Cultural differences in attitudes towards robots.*

Baskerville, R. F. (2003). Hofstede never studied culture. *Accounting, Organizations and Society, 28*(1), 1–14. https://doi.org/10.1016/S0361-3682(01)00048-4

Baughan, A., & Reinecke, K. (2021). Do cross-cultural differences in visual attention patterns affect search efficiency on websites? In *Conference on Human Factors in Computing Systems - Proceedings.*

Baughan, A., Oliveira, N., August, T., Yamashita, N., & Reinecke, K. (2021). Do cross-cultural differences in visual attention patterns affect search efficiency on websites? In *Proceedings of the 2021 CHI Conference on Human Factors in Computing Systems* (pp. 1–12). https://doi.org/10.1145/3411764.3445519

Bell, D. (1997). *Consuming geographies: We are where we eat.* Routledge.

Bell, D. (2002). *An introduction to cybercultures.* Routledge.

Bell, G. (2006). The age of the thumb: A cultural reading of mobile technologies from Asia. *Knowledge, Technology, & Policy, 19*(2), 41–57. https://doi.org/10.14361/9783839404034-004

Benaida, M. (2022). Significance of culture toward the usability of web design and its relationship with satisfaction. *Universal Access in the Information Society, 21*(3), 625–638. https://doi.org/10.1007/s10209-021-00799-y

Benford, S., Greenhalgh, C., Anderson, B., Jacobs, R., Golembewski, M., Jirotka, M., Stahl, B. C., Timmermans, J., Giannachi, G., Adams, M., Farr, J. R., Tandavanitj, N., & Jennings, K. (2015). The ethical implications of HCI's turn to the cultural. *ACM Transactions on Computer-Human Interaction, 22*(5), 1–37. https://doi.org/10.1145/2775107

Benjamin, R. (2019). *Race after technology. Abolitionist tools for the new Jim code*. Polity.

Bennett, T., Grossberg, L., Morris, M., & Williams, R. (Eds.). (2005). *New keywords: A revised vocabulary of culture and society*. Blackwell Pub.

Beu, A., Honold, P., & Yuan, X. (n.d.). *How to build up an infrastructure for intercultural usability engineering*.

Beyer, H. (2010). *User-centered agile methods*. Morgan & Claypool Publishers.

Bijker, W. (1995). *Of bicycles, bakelites, and bulbs*. The MIT Press.

Blume, S. S. (2010). *The artificial ear: Cochlear implants and the culture of deafness*. Rutgers University Press.

Bobeth, J., Schreitter, S., Schmehl, S., Deutsch, S., & Tscheligi, M. (2013). User-centered design between cultures: Designing for and with immigrants. In P. Kotzé, G. Marsden, G. Lindgaard, J. Wesson, & M. Winckler (Eds.), *Human-computer interaction – INTERACT 2013* (Vol. 8120, pp. 713–720). Springer. https://doi.org/10.1007/978-3-642-40498-6_65

Bødker, S. (2006). When second wave HCI meets third wave challenges. NordiCHI, 14–18 October.

Boehner, K., Vertesi, J., Sengers, P., & Dourish, P. (2007). How HCI interprets the probes. In *Proceedings of the SIGCHI Conference on Human Factors in Computing Systems* (pp. 1077–1086). https://doi.org/10.1145/1240624.1240789

Böhm, V., & Wolff, C. (2014). A review of empirical intercultural usability studies. In A. Marcus (Ed.), *Design, user experience, and usability. Theories, methods, and tools for designing the user experience* (Vol. 8517, pp. 14–24). Springer International Publishing. https://doi.org/10.1007/978-3-319-07668-3_2

Boujarwah, F. A., Nazneen, Hong, H., Abowd, G. D., & Arriaga, R. I. (2011). Towards a framework to situate assistive technology design in the context of culture. In *The Proceedings of the 13th International ACM SIGACCESS Conference on Computers and Accessibility* (pp. 19–26). https://doi.org/10.1145/2049536.2049542

Bourges-Waldegg, P., & Scrivener, S. A. R. (1998). Meaning, the central issue in cross-cultural HCI design. *Interacting with Computers, 9*(3), 287–309. https://doi.org/10.1016/S0953-5438(97)00032-5

Boutyline, A., & Soter, L. K. (n.d.). Cultural schemas: What they are, how to find them, and what to do once you've caught one. *American Sociological Review*.

Bragg, D., Koller, O., Bellard, M., Berke, L., Boudreault, P., Braffort, A., Caselli, N., Huenerfauth, M., Kacorri, H., Verhoef, T., Vogler, C., & Ringel Morris, M. (2019). Sign language recognition, generation, and translation: An interdisciplinary perspective. In *The 21st International ACM SIGACCESS Conference on Computers and Accessibility* (pp. 16–31). https://doi.org/10.1145/3308561.3353774

Braybrooke, K., Janes, S., & Sato, C. (2021). Care-full design sprints, online? Addressing gaps in cultural access and inclusion during Covid-19 with vulnerable communities in London and Tokyo. In *C&T'21: Proceedings of the 10th International Conference on Communities & Technologies - Wicked Problems in the Age of Tech* (pp. 25–37). https://doi.org/10.1145/3461564.3461583

Breazeal, C., Dautenhahn, K., & Kanda, T. (2016). Social robotics. In B. Siciliano & O. Khatib (Eds.), *Springer handbook of robotics* (pp. 1935–1961). Springer.

Brianza, G., Benjamin, J., Cornelio, P., Maggioni, E., & Obrist, M. (2022). QuintEssence: A probe study to explore the power of smell on emotions, memories, and body image in daily life. *ACM Transactions on Computer-Human Interaction, 29*(6), 1–33. https://doi.org/10.1145/3526950

Bryan-Kinns, N., Wang, W., & Ji, T. (2022). Qi2He: A co-design framework inspired by eastern epistemology. *International Journal of Human-Computer Studies, 160*, 102773. https://doi.org/10.1016/j.ijhcs.2022.102773

Buchtel, E. E. (2014). Cultural sensitivity or cultural stereotyping? Positive and negative effects of a cultural psychology class. *International Journal of Intercultural Relations, 39*, 40–52. https://doi.org/10.1016/j.ijintrel.2013.09.003

Bucolo, S. (2004). Understanding cross cultural differences during interaction within immersive virtual environments. In *Proceedings of the 2004 ACM SIGGRAPH International Conference on Virtual Reality Continuum and Its Applications in Industry - VRCAI'04* (p. 221). https://doi.org/10.1145/1044588.1044634

Bühler, D., Hemmert, F., Hurtienne, J., & Petersen, C. (2022). Designing universal and intuitive pictograms (UIPP)—A detailed process for more suitable visual representations. *International Journal of Human-Computer Studies, 163*, 102816. https://doi.org/10.1016/j.ijhcs.2022.102816

Burke, P. (2009). *Cultural hybridity*. Polity Press.

Burrows, A., Mitchell, V., & Nicolle, C. (2015). Cultural probes and levels of creativity. In *Proceedings of the 17th International Conference on Human-Computer Interaction with Mobile Devices and Services Adjunct* (pp. 920–923). https://doi.org/10.1145/2786567.2794302

Cabrero, D. G., Winschiers-Theophilus, H., & Abdelnour-Nocera, J. (2016). A critique of personas as representations of "the other" in cross-cultural technology design. In *Proceedings of the First African Conference on Human Computer Interaction* (pp. 149–154). https://doi.org/10.1145/2998581.2998595

Callahan, E. (2006). Interface design and culture. *Annual Review of Information Science and Technology, 39*(1), 255–310. https://doi.org/10.1002/aris.1440390114

Cannanure, V. K., Karusala, N., Rivera-Loaiza, C., Prabhakar, A. S., Varanasi, R. A., Tuli, A., Gamage, D., Noor, F., Nemer, D., Das, D., Dray, S., Sturm, C., & Kumar, N. (2022a). HCI across borders: Navigating shifting borders at CHI. In *CHI Conference on Human Factors in Computing Systems Extended Abstracts* (pp. 1–5). https://doi.org/10.1145/3491101.3503706

Cannanure, V. K., Nemer, D., & Sturm, C. (2022b). HCI across borders: Navigating shifting borders at CHI. In *Conference on Human Factors in Computing Systems – Proceedings*.

Card, S., Moran, T., & Newell, A. (1983). *The psychology of human-computer interaction*. Lawrence Erlbaum Associates.

Cardoso, P. J. S., Rodrigues, J. M. F., Pereira, J., Nogin, S., Lessa, J., Ramos, C. M. Q., Bajireanu, R., Gomes, M., & Bica, P. (2020). Cultural heritage visits supported on visitors' preferences and mobile devices. *Universal Access in the Information Society, 19*(3), 499–513. https://doi.org/10.1007/s10209-019-00657-y

Carroll-Miranda, J., Ordonez, P., Orozco, E., Bravo, M., Borrero, M., Lopez, L., Houser, G., Gerena, E., Reed, D., Santiago, B., Corchado, A., & Santana, A. (2019). This is what diversity looks like: Making CS curriculum culturally relevant for Spanish-speaking communities. In *Proceedings of the 50th ACM Technical Symposium on Computer Science Education* (pp. 647–648). https://doi.org/10.1145/3287324.3287339

Cave, S., & Dihal, K. (2020). The whiteness of AI. *Philosophy & Technology, 33*, 685–703. https://doi.org/10.1007/s13347-020-00415-6

Cesário, V., Coelho, A., & Nisi, V. (2019). "This Is Nice but That Is Childish": Teenagers evaluate museum-based digital experiences developed by cultural heritage professionals. In *Extended Abstracts of the Annual Symposium on Computer-Human Interaction in Play Companion Extended Abstracts* (pp. 159–169). https://doi.org/10.1145/3341215.3354643

Chavan, A. L. (2005). *Another culture, another method evaluation Bollywood style*. Human Factors.

Chen, A., Mashhadi, A., Ang, D., & Harkrider, N. (1999). Cultural issues in the design of technology-enhanced learning systems. *British Journal of Educational Technology, 30*(3), 217–230. https://doi.org/10.1111/1467-8535.00111

Chen, C., Garrod, O. G. B., Ince, R. A. A., Foster, M. E., Schyns, P. G., & Jack, R. E. (2020). Building culturally-valid dynamic facial expressions for a conversational virtual agent using human perception. In *Proceedings of the 20th ACM International Conference on Intelligent Virtual Agents* (pp. 1–3). https://doi.org/10.1145/3383652.3423913

Chiao, J. Y. (2009). Cultural neuroscience: A once and future discipline. In *Progress in brain research* (Vol. 178, pp. 287–304). Elsevier. https://doi.org/10.1016/S0079-6123(09)17821-4

Chien, S.-Y., Lewis, M., Sycara, K., Kumru, A., & Liu, J.-S. (2020). Influence of culture, transparency, trust, and degree of automation on automation use. *IEEE Transactions on Human-Machine Systems, 50*(3), 205–214. https://doi.org/10.1109/THMS.2019.2931755

Chiotaki, D., & Karpouzis, K. (2020). Open and cultural data games for learning. In *International Conference on the Foundations of Digital Games* (pp. 1–7). https://doi.org/10.1145/3402942.3409621

Chizari, S. (2016). Exploring the role of culture in online searching behavior from cultural cognition perspective. In *Proceedings of the 2016 ACM on Conference on Human Information Interaction and Retrieval* (pp. 349–351). https://doi.org/10.1145/2854946.2854954

Cho, H., Knijnenburg, B., Kobsa, A., & Li, Y. (2018). Collective privacy management in social media: A cross-cultural validation. *ACM Transactions on Computer-Human Interaction, 25*(3), 1–33. https://doi.org/10.1145/3193120

Choi, B., Lee, I., Kim, J., & Jeon, Y. (2005a). A qualitative cross-national study of cultural influences on mobile data service design. In *Proceedings of the SIGCHI Conference on Human Factors in Computing Systems* (pp. 661–670). https://doi.org/10.1145/1054972.1055064

Chu, J. H. (2016). Design space for tangible and embodied interaction with cultural heritage. In *Proceedings of the 2016 ACM Conference Companion Publication on Designing Interactive Systems* (pp. 27–28). https://doi.org/10.1145/2908805.2909420

Ciolfi, L., Damala, A., Hornecker, E., Lechner, M., Maye, L., & Petrelli, D. (2015). Cultural heritage communities: Technologies and challenges. In *Proceedings of the 7th International Conference on Communities and Technologies* (pp. 149–152). https://doi.org/10.1145/2768545.2768560

Clemmensen, T. (2011). Templates for cross-cultural and culturally specific usability testing: Results from field studies and ethnographic interviewing in three countries. *International Journal of Human-Computer Interaction, 27*(7), 634–669. https://doi.org/10.1080/10447318.2011.555303

Clemmensen, T., & Roese, K. (2010a). An overview of a decade of journal publications about culture and human-computer interaction (HCI). *IFIP Advances in Information and Communication Technology, 316*, 98–112. https://doi.org/10.1007/978-3-642-11762-6_9

Clemmensen, T., & Roese, K. (2010b). An overview of a decade of journal publications about culture and human-computer interaction (HCI). In D. Katre, R. Orngreen, P. Yammiyavar, & T. Clemmensen (Eds.), *Human work interaction design: Usability in social, cultural and organizational contexts* (Vol. 316, pp. 98–112). Springer. https://doi.org/10.1007/978-3-642-11762-6_9

Clemmensen, T., Hertzum, M., Hornbæk, K., Shi, Q., & Yammiyavar, P. (2009). Cultural cognition in usability evaluation. *Interacting with Computers, 21*(3), 212–220. https://doi.org/10.1016/j.intcom.2009.05.003

Cockton, G. (2005). A development framework for value-centred design. In *CHI'05 Extended Abstracts on Human Factors in Computing Systems* (pp. 1292–1295). https://doi.org/10.1145/1056808.1056899

Cooney, S. (2021). Riding the bus in Los Angeles: Creating cultural micro-exposures via technology. In *Extended Abstracts of the 2021 CHI Conference on Human Factors in Computing Systems* (pp. 1–9). https://doi.org/10.1145/3411763.3450392

Corrêa, J. M., & Mota, M. P. (2021). Teaching programming using cultural viewpoint metaphors resignification. In *X Latin American Conference on Human Computer Interaction* (pp. 1–5). https://doi.org/10.1145/3488392.3488395

Cramtom, C. D., & Hinds, P. J. (2007). Intercultural interaction in distributed teams: Salience of and adaptations to cultural differences. *Academy of Management Proceedings, 2007*(1), 1–6. https://doi.org/10.5465/ambpp.2007.26523095

Cui, C. (2016). A study of digital games as a new media of cultural transmission. In Z. Pan, A. D. Cheok, W. Müller, & M. Zhang (Eds.), *Transactions on edutainment XII* (Vol. 9292, pp. 48–52). Springer. https://doi.org/10.1007/978-3-662-50544-1_4

Cummings, D., Cheeks, L., & Robinson, R. (2018). Culturally-centric outreach and engagement for underserved groups in STEM. In *Proceedings of the 49th ACM Technical Symposium on Computer Science Education* (pp. 447–452). https://doi.org/10.1145/3159450.3159565

Cyr, D., Head, M., & Larios, H. (2010). Colour appeal in website design within and across cultures: A multi-method evaluation. *International Journal of Human-Computer Studies, 68*(1–2), 1–21. https://doi.org/10.1016/j.ijhcs.2009.08.005

Das Swain, V., Saha, K., Reddy, M. D., Rajvanshy, H., Abowd, G. D., & De Choudhury, M. (2020). Modeling organizational culture with workplace experiences shared on glassdoor. In *Proceedings of the 2020 CHI Conference on Human Factors in Computing Systems* (pp. 1–15). https://doi.org/10.1145/3313831.3376793

Davis, J., Lachney, M., Zatz, Z., Babbitt, W., & Eglash, R. (2019). A cultural computing curriculum. In *Proceedings of the 50th ACM Technical Symposium on Computer Science Education* (pp. 1171–1175). https://doi.org/10.1145/3287324.3287439

de Mooij, M. (2011). *Consumer behavior and culture. Consequences for global marketing and advertising*. SAGE.

de Souza, T. R. C. B., & Bernardes, J. L. (2016). The influences of culture on user experience. In P.-L. P. Rau (Ed.), *Cross-cultural design* (Vol. 9741, pp. 43–52). Springer International Publishing. https://doi.org/10.1007/978-3-319-40093-8_5

Del Galdo, E. M., & Nielsen, J. (Eds.). (1996). *International user interfaces*. Wiley.

Dell, N., Vaidyanathan, V., Medhi, I., Cutrell, E., & Thies, W. (2012). "Yours is better!": Participant response bias in HCI. In *Proceedings of the SIGCHI Conference on Human Factors in Computing Systems* (pp. 1321–1330). https://doi.org/10.1145/2207676.2208589

Dhaundiyal, D., Chakravarty, R., & N. Joshi, A. (2020). Hofstede and Hobbitses: Generational evolution of power distance and masculinity in UK in popular literature. In *IndiaHCI '20: Proceedings of the 11th Indian Conference on Human-Computer Interaction* (pp. 1–11). https://doi.org/10.1145/3429290.3429291

Diamant, E. I., Fussell, S. R., & Lo, F. (n.d.). *Where did we turn wrong? Unpacking the effects of culture and technology on attributions of team performance.*

Ding, K.-Y., Huang, H.-H., Berberich, N., Kaseda, M., Kuwabara, K., & Nishida, T. (2019). Designing a data corpus of collaborative group tasks with the members from unbalanced cultural backgrounds. In *Proceedings of the 7th International Conference on Human-Agent Interaction* (pp. 321–323). https://doi.org/10.1145/3349537.3352807

Dix, A., Finlay, J., Abowd, G., & Beale, R. (2004). *Human-computer interaction*. Pearson Education.

Dogra, N., Reitmanova, S., & Carter-Pokras, O. (2010). Teaching cultural diversity: Current status in U.K., U.S., and Canadian medical schools. *Journal of General Internal Medicine, 25*(S2), 164–168. https://doi.org/10.1007/s11606-009-1202-7

Dong, Y., & Lee, K. P. (2008). A cross-cultural comparative study of users' perceptions of a web-page: With a focus on the cognitive styles of Chinese, Koreans and Americans. *International Journal of Design, 2*(2), 19–30.

Douglas, I., & Liu, Z. (Eds.). (2011). *Global usability.* Springer. https://doi.org/10.1007/978-0-85729-304-6

Dourish, P. (2004). What we talk about when we talk about context. *Personal and Ubiquitous Computing, 8*(1), 19–30. https://doi.org/10.1007/s00779-003-0253-8

Dourish, P. (2016). Algorithms and their others: Algorithmic culture in context. *Big Data & Society, 3*(2), 205395171666512. https://doi.org/10.1177/2053951716665128

Dreschler-Fischer, D. L. (n.d.). *Prof. Dr. Christiane Floyd Prof. Dr. Leonie Dreschler-Fischer Dr. Edla Faust Ramos.*

du Gay, P., Hall, S., Janes, L., Mackay, H., & Negus, K. (2003). *Doing cultural studies. The story of the Sony Walkman.* SAGE.

Duan, W., & Fussell, S. R. (2021). Understanding and identifying design opportunities for facilitating humorous interactions in multilingual multicultural contexts. In *Extended Abstracts of the 2021 CHI Conference on Human Factors in Computing Systems* (pp. 1–6). https://doi.org/10.1145/3411763.3451668

Duarte, A. M. B., Brendel, N., Degbelo, A., & Kray, C. (2018). Participatory design and participatory research: An HCI case study with young forced migrants. *ACM Transactions on Computer-Human Interaction, 25*(1), 1–39. https://doi.org/10.1145/3145472

Duncker, E. (2002). Cross-cultural usability of the library metaphor. In *Proceedings of the 2nd ACM/IEEE-CS Joint Conference on Digital Libraries* (pp. 223–230). https://doi.org/10.1145/544220.544269

Eckhardt, J., Kaletka, C., Pelka, B., Unterfrauner, E., Voigt, C., & Zirngiebl, M. (2021). Gender in the making: An empirical approach to understand gender relations in the maker movement. *International Journal of Human-Computer Studies, 145*, 102548. https://doi.org/10.1016/j.ijhcs.2020.102548

Editorial. (2001). *Disability culture* - Independent Living Institute Newsletter.

Egan, C., & Benyon, D. (2017). Sustainable HCI: Blending permaculture and user-experience. In *Proceedings of the 2017 ACM Conference Companion Publication on Designing Interactive Systems* (pp. 39–43). https://doi.org/10.1145/3064857.3079115

Eglash, R., Gilbert, J. E., & Foster, E. (2013). Toward culturally responsive computing education. *Communications of the ACM, 56*(7), 33–36. https://doi.org/10.1145/2483852.2483864

Ekman, P. (2003). *Emotions revealed.* Times Books.

Ekuan, K. (2000). *The aesthetics of the Japanese Lunchbox.* The MIT Press.

Elliott, A. (2019). *The culture of AI: Everyday life and the digital revolution* (1st ed.). Routledge. https://doi.org/10.4324/9781315387185

Endrass, B., Andr, E., Rehm, M., Lipi, A. A., & Nakano, Y. (n.d.). *Culture-related differences in aspects of behavior for virtual characters across Germany and Japan.*

Endrass, B., André, E., Rehm, M., & Nakano, Y. (2013). Investigating culture-related aspects of behavior for virtual characters. *Autonomous Agents and Multi-Agent Systems, 27*(2), 277–304. https://doi.org/10.1007/s10458-012-9218-5

English-Lueck, J. A. (2017). *Cultures@SiliconValley (Second edition).* Stanford University Press.

Epp, F. A., Kantosalo, A., Jain, N., Lucero, A., & Mekler, E. D. (2022). Adorned in memes: Exploring the adoption of social wearables in Nordic student culture. In *CHI Conference on Human Factors in Computing Systems* (pp. 1–18). https://doi.org/10.1145/3491102.3517733

Escobedo, L., & Arriaga, R. I. (2022). Understanding the challenges of deploying a milestone-tracking application in a cross-cultural context. *International Journal of Human-Computer Studies, 162,* 102801. https://doi.org/10.1016/j.ijhcs.2022.102801

Evers, V., & Day, D. (1997). The role of culture in interface acceptance. In S. Howard, J. Hammond, & G. Lindgaard (Eds.), *Human-computer interaction INTERACT'97* (pp. 260–267). Springer. https://doi.org/10.1007/978-0-387-35175-9_44

Faisal, C. M. N., Gonzalez-Rodriguez, M., Fernandez-Lanvin, D., & de Andres-Suarez, J. (2017). Web design attributes in building user trust, satisfaction, and loyalty for a high uncertainty avoidance culture. *IEEE Transactions on Human-Machine Systems, 47*(6), 847–859. https://doi.org/10.1109/THMS.2016.2620901

Fallatah, A., Chun, B., Balali, S., & Knight, H. (2020). "Would You Please Buy Me a Coffee?": How microcultures impact people's helpful actions toward robots. In *Proceedings of the 2020 ACM Designing Interactive Systems Conference* (pp. 939–950). https://doi.org/10.1145/3357236.3395446

Fang, T., & Zhao, F. (2021). Research on digital dissemination of Chinese classical garden culture. In M. Rauterberg (Ed.), *Culture and computing. Interactive cultural heritage and arts* (Vol. 12794, pp. 63–73). Springer International Publishing. https://doi.org/10.1007/978-3-030-77411-0_5

Faucher, C. (Ed.). (2018). *Advances in culturally-aware intelligent systems and in cross-cultural psychological studies* (Vol. 134). Springer International Publishing. https://doi.org/10.1007/978-3-319-67024-9

Feinberg, M., Fox, S., Hardy, J., Steinhardt, S., & Vaghela, P. (2019). At the intersection of culture and method: Designing feminist action. In *Companion Publication of the 2019 on Designing Interactive Systems Conference 2019 Companion* (pp. 365–368). https://doi.org/10.1145/3301019.3319993

Fernandes, T. (1994). Global interface design. In *Conference Companion on Human Factors in Computing Systems - CHI'94* (pp. 373–374). https://doi.org/10.1145/259963.260509

Ferwerda, B., & Bauer, C. (2022). To flip or not to flip: Conformity effect across cultures. In *CHI Conference on Human Factors in Computing Systems Extended Abstracts* (pp. 1–7). https://doi.org/10.1145/3491101.3519662

Fetterman, D. (2010). *Ethnography. Step-by-Step.* SAGE.

Ford, G., & Kotzé, P. (2005). Designing usable interfaces with cultural dimensions. In M. F. Costabile & F. Paternò (Eds.), *Human-computer interaction—INTERACT 2005* (Vol. 3585, pp. 713–726). Springer. https://doi.org/10.1007/11555261_57

Frandsen-Thorlacius, O., Hornbæk, K., Hertzum, M., & Clemmensen, T. (2009). Non-universal usability?: A survey of how usability is understood by Chinese and Danish users. In *Proceedings of the SIGCHI Conference on Human Factors in Computing Systems* (pp. 41–50). https://doi.org/10.1145/1518701.1518708

Franklin, D., Weintrop, D., Palmer, J., Coenraad, M., Cobian, M., Beck, K., Rasmussen, A., Krause, S., White, M., Anaya, M., & Crenshaw, Z. (2020). Scratch encore: The design and pilot of a culturally-relevant intermediate scratch curriculum. In *Proceedings of the 51st ACM Technical Symposium on Computer Science Education* (pp. 794–800). https://doi.org/10.1145/3328778.3366912

Friedman, B., Hook, K., Gill, B., Eidmar, L., Prien, C. S., & Severson, R. (2008). Personlig integritet: A comparative study of perceptions of privacy in public places in Sweden and the United States. In *Proceedings of the 5th Nordic Conference on Human-Computer Interaction: Building Bridges* (pp. 142–151). https://doi.org/10.1145/1463160.1463176

Friedman, B., Jr, P. H. K., Hagman, J., Severson, R. L., & Gill, B. (n.d.). *The watcher and the watched: Social judgments about privacy in a public place.*

Friedman, T. (2005). *The world if flat. A brief history of the 21st century.*

Garcia-Gavilanes, R., Quercia, D., & Jaimes, A. (2021). Cultural dimensions in Twitter: Time, individualism and power. *Proceedings of the International AAAI Conference on Web and Social Media, 7*(1), 195–204. https://doi.org/10.1609/icwsm.v7i1.14419

Garcia, R., & Poblete, B. (n.d.). *Microblogging without borders: Differences and similarities.*

Gaver, B., Dunne, T., & Pacenti, E. (1999). Design: Cultural probes. *Interactions, 6*(1), 21–29. https://doi.org/10.1145/291224.291235

Gay, G. (2018). *Culturally responsive teaching. Theory, research, and practice.* Teachers College Press.

Gayler, T., Sas, C., & Kalnikaitė, V. (2022). Exploring the design space for human-food-technology interaction: An approach from the lens of eating experiences. *ACM Transactions on Computer-Human Interaction, 29*(2), 1–52. https://doi.org/10.1145/3484439

Geertz, C. (1973). *The interpretation of cultures.* Basic Books.

Geertz, C. (1983). *Local knowledge. Further essays in interpretive anthropology.* Basic Books. https://doi.org/10.1177/1350507602334002

George, R., Nesbitt, K., Gillard, P., & Donovan, M. (2010). Identifying cultural design requirements for an Australian indigenous website. *User Interfaces, 106.*

Gibbons, J., & Poelker, K. (2019). Adolescent development in a cross-cultural perspective. In Keith, K. (Ed.), *Cross-cultural psychology. Contemporary themes and perspectives* (pp. 190–216). Wiley, Blackwell.

Giglitto, D., Ciolfi, L., Claisse, C., & Lockley, E. (2019). Bridging cultural heritage and communities through digital technologies: Understanding perspectives and challenges. In *Proceedings of the 9th International Conference on Communities & Technologies - Transforming Communities* (pp. 81–91). https://doi.org/10.1145/3328320.3328386

Gou, Y. (2021). Computer digital technology in the design of intangible cultural heritage protection platform. In *2021 3rd International Conference on Artificial Intelligence and Advanced Manufacture* (pp. 1524–1528). https://doi.org/10.1145/3495018.3495434

Goulding, A., Campbell-Meier, J., & Sylvester, A. (2020). Indigenous cultural sustainability in a digital world: Two case studies from Aotearoa New Zealand. In A. Sundqvist, G. Berget, J. Nolin, & K. I. Skjerdingstad (Eds.), *Sustainable digital communities* (Vol. 12051, pp. 66–75). Springer International Publishing. https://doi.org/10.1007/978-3-030-43687-2_5

Goyal, N., Miner, W., & Nawathe, N. (2012). Cultural differences across governmental website design. In *Proceedings of the 4th International Conference on Intercultural Collaboration* (pp. 149–152). https://doi.org/10.1145/2160881.2160907

Greenbaum, J. M., & Kyng, M. (Eds.). (1991). *Design at work: Cooperative design of computer systems.* L. Erlbaum Associates.

Groh, A. (2020). *Theories of culture.* Routledge, Taylor & Francis Group.

Grudin, J. (2012). A moving target: The evolution of human-computer interaction. In J. Jacko (Ed.), *The human-computer interaction handbook. Fundamentals, evolving technologies, and emerging applications.* CRC Press.

Grudin, J. (2017). From tool to partner: The evolution of human-computer interaction. In *Synthesis lectures on human-centered informatics* (Vol. 10, Issue 1). Morgan and Claypool. https://doi.org/10.2200/S00745ED1V01Y201612HCI035

Gudykunst, W. B., Matsumoto, Y., Ting-Toomey, S., Nishida, T., Kim, K., & Heyman, S. (1996). The influence of cultural individualism-collectivism, self construals, and individual values on communication styles across cultures. *Human Communication Research, 22*(4), 510–543. https://doi.org/10.1111/j.1468-2958.1996.tb00377.x

Guo, J., Lin, Y., Yang, H., Wang, J., Li, S., Liu, E., Yao, C., & Ying, F. (2020). Comparing the tangible tutorial system and the human teacher in intangible cultural heritage education. In *Proceedings of the 2020 ACM Designing Interactive Systems Conference* (pp. 895–907). https://doi.org/10.1145/3357236.3395449

Hakken, D., & Maté, P. (2014). The culture question in participatory design. In *Proceedings of the 13th Participatory Design Conference: Short Papers, Industry Cases, Workshop Descriptions, Doctoral Consortium Papers, and Keynote Abstracts* (Vol. 2, pp. 87–91). https://doi.org/10.1145/2662155.2662197

Häkkilä, J., Wiberg, M., Eira, N. J., Seppänen, T., Juuso, I., Mäkikalli, M., & Wolf, K. (2020). Design sensibilities—designing for cultural sensitivity. In *Proceedings of the 11th Nordic Conference on Human-Computer Interaction: Shaping Experiences, Shaping Society* (pp. 1–3). https://doi.org/10.1145/3419249.3420100

Halabi, A., & Zimmermann, B. (2019). Waves and forms: Constructing the cultural in design. *AI & Society, 34*(3), 403–417. https://doi.org/10.1007/s00146-017-0713-8

Hall, E. (1973). *The silent language.* Anchor Books.

Hall, E. (1976). *Beyond culture.* Anchor Books.

Hall, E. T. (1990). *The hidden dimension.* Anchor Books.

Hall, M., Jong, M. D., & Steehouder, M. L. (n.d.). *Individualistic and collectivistic participants compared.*

Hao, C., Van Boeijen, A., & Stappers, P. J. (2017). Cultura: A communication toolkit for designers to gain empathic insights across cultural boundaries. In *Proceedings of the IASDR Conference RE: Research* (pp. 497–510). https://doi.org/10.7945/C2SD5J

Harper, R., Rodden, T., Rogers, Y., & Sellen, A. (Eds.). (2007). *Being human: Human computer interaction in the year 2020.* https://doi.org/10.1145/1467247.1467265

Harrell, D. F., Vieweg, S., Kwak, H., Lim, C.-U., Sengun, S., Jahanian, A., & Ortiz, P. (2017). Culturally-grounded analysis of everyday creativity in social media: A case study in Qatari context. In *Proceedings of the 2017 ACM SIGCHI Conference on Creativity and Cognition* (pp. 209–221). https://doi.org/10.1145/3059454.3059456

Harrison, S., Sengers, P., & Tatar, D. (2011). Making epistemological trouble: Third- paradigm HCI as successor science. *Interacting with Computers, 23*(5), 385–392. https://doi.org/10.1016/j.intcom.2011.03.005

Hassenzahl, M. (2010). *Experience design.* Springer.

Hatch, M. (2014). *The maker movement manifesto: Rules for innovation in the new world of crafters, hackers, and tinkerers.* McGrawHill.

Haviland, W., Prins, H., McBride, B., & Walrath, D. (2011). *Cultural anthropology.* Wadsworth.

He, H. A., Memarovic, N., Sabiescu, A., & de Moor, A. (2015). CulTech2015: Cultural diversity and technology design. In *Proceedings of the 7th International Conference on Communities and Technologies* (pp. 153–156). https://doi.org/10.1145/2768545.2768561

Hebdige, D. (1991). *Subculture: The meaning of style.* Routledge.

Heimgärtner, R. (2013). Intercultural user interface design — culture-centered HCI design — cross-cultural user interface design: Different terminology or different approaches? In A. Marcus (Ed.), *Design, user experience, and usability. Health, learning, playing, cultural, and cross-cultural user experience* (Vol. 8013, pp. 62–71). Springer. https://doi.org/10.1007/978-3-642-39241-2_8

Heimgärtner, R. (2019). *Intercultural user interface design*. Springer International Publishing. https://doi.org/10.1007/978-3-030-17427-9

Heimgärtner, R., Solanki, A., & Windl, H. (2017). Cultural user experience in the car—toward a standardized systematic intercultural agile automotive UI/UX design process. In G. Meixner & C. Müller (Eds.), *Automotive user interfaces* (pp. 143–184). Springer International Publishing. https://doi.org/10.1007/978-3-319-49448-7_6

Heine, S. J. (2016). *Cultural psychology* (3rd ed.). Norton & Company Inc.

Heintze, K. E., Krämer, N., Foster, D., & Lawson, S. (2015). Designing student energy interventions: A cross-cultural comparison. In *Proceedings of the 2015 British HCI Conference* (pp. 247–254). https://doi.org/10.1145/2783446.2783575

Hergeth, S., Lorenz, L., Krems, J. F., & Toenert, L. (2015). Effects of take-over requests and cultural background on automation trust in highly automated driving. In *Proceedings of the 8th International Driving Symposium on Human Factors in Driver Assessment, Training, and Vehicle Design: Driving Assessment 2015* (pp. 331–337). https://doi.org/10.17077/drivingassessment.1591

Herman, L. (1996). Towards effective usability evaluation in Asia: Cross-cultural differences. In *Proceedings Sixth Australian Conference on Computer-Human Interaction* (pp. 135–136). https://doi.org/10.1109/OZCHI.1996.559999

Hertzum, M. (2010). Images of usability. *International Journal of Human-Computer Interaction, 26*(6), 567–600. https://doi.org/10.1080/10447311003781300

Hertzum, M. (2018). Commentary: Usability—a sensitizing concept. *Human-Computer Interaction, 33*(2), 178–181. https://doi.org/10.1080/07370024.2017.1302800

Hertzum, M., Clemmensen, T., Hornbæk, K., Kumar, J., Shi, Q., & Yammiyavar, P. (2007). Usability constructs: A cross-cultural study of how users and developers experience their use of information systems. In N. Aykin (Ed.), *Usability and internationalization. HCI and culture* (Vol. 4559, pp. 317–326). Springer. https://doi.org/10.1007/978-3-540-73287-7_39

Hestres, L. (n.d.). *The influence of American culture on software design: Microsoft Outlook as a case study*.

Hodgson, A., Siemieniuch, C. E., & Hubbard, E.-M. (2013). Culture and the safety of complex automated sociotechnical systems. *IEEE Transactions on Human-Machine Systems, 43*(6), 608–619. https://doi.org/10.1109/THMS.2013.2285048

Hodkinson, P., & Deicke, W. (Eds.). (2007). *Youth cultures: Scenes, subcultures and tribes*. Routledge.

Hofstede, G. (2002). *Exploring culture. Exercises, stories, and synthetic cultures*. Intercultural Press.

Hofstede, G., Hofstede, G. J., & Minkov, M. (2010). *Cultures and organizations*. McGraw Hill.

Holmes, N. (2011). The cultural potential of keyboards. *Computer, 44*(8), 112–111. https://doi.org/10.1109/MC.2011.255

Honeywood, R., & Fung, J. (2012). Best practices for game localization. In *IGDA Game Localization SIG*. http://www.igda.org/members/group_content_view.asp?group=121045&id=526230

Honold, P. (2000). Culture and context: An empirical study for the development of a framework for the elicitation of cultural influence in product usage. *International Journal of Human-Computer Interaction, 12*(3–4), 327–345. https://doi.org/10.1080/10447318.2000.9669062

Horn, M. S. (2013). The role of cultural forms in tangible interaction design. In *Proceedings of the 7th International Conference on Tangible, Embedded and Embodied Interaction* (pp. 117–124). https://doi.org/10.1145/2460625.2460643

Horn, M. S., AlSulaiman, S., & Koh, J. (2013). Translating Roberto to Omar: Computational literacy, stickerbooks, and cultural forms. In *Proceedings of the 12th International Conference on Interaction Design and Children* (pp. 120–127). https://doi.org/10.1145/2485760.2485773

Hu, X., & Yang, Y.-H. (2017). Cross-dataset and cross-cultural music mood prediction: A case on Western and Chinese pop songs. *IEEE Transactions on Affective Computing, 8*(2), 228–240. https://doi.org/10.1109/TAFFC.2016.2523503

Huang, H.-H., Tarasenko, K., Nishida, T., Cerekovic, A., Levacic, V., Zoric, G., Pandzic, I. S., & Nakano, Y. (2007). An agent based multicultural tour guide system with nonverbal user interface. *Journal on Multimodal User Interfaces, 1*(1), 41–48. https://doi.org/10.1007/BF02884431

Hutchins, E. (1995). *Cognition in the wild*. The MIT Press.

Hwang, I. D., Guglielmetti, M., & Dziekan, V. (2015). Super-natural: Digital life in eastern culture. *SIGGRAPH ASIA 2015 Art Papers*, 1–7. https://doi.org/10.1145/2835641.2835644

Investigating A Culture of Disability: Final Report. (n.d.).

Irani, L. (2010). HCI on the move: Methods, culture, values. In *CHI'10 Extended Abstracts on Human Factors in Computing Systems* (pp. 2939–2942). https://doi.org/10.1145/1753846.175 3890

Irani, L., Vertesi, J., Dourish, P., Philip, K., & Grinter, R. E. (2010). Postcolonial computing: A lens on design and development. In *Proceedings of the SIGCHI Conference on Human Factors in Computing Systems* (pp. 1311–1320). https://doi.org/10.1145/1753326.1753522

Isbister, K., Nakanishi, H., Ishida, T., & Nass, C. (2000). Helper agent: Designing an assistant for human-human interaction in a virtual meeting space. In *Proceedings of the SIGCHI Conference on Human Factors in Computing Systems* (pp. 57–64). https://doi.org/10.1145/332040.332407

Ishak, Z., Jaafar, A., & Nayan, N. M. (2015). Methodology for the development of interface design guidelines based on local cultural dimensions. In C. Stephanidis (Ed.), *HCI International 2015— Posters' Extended Abstracts* (Vol. 528, pp. 245–248). Springer International Publishing. https://doi.org/10.1007/978-3-319-21380-4_43

Ishioh, T., & Koda, T. (2016a). Cross-cultural study of perception and acceptance of Japanese self-adaptors. In *Proceedings of the Fourth International Conference on Human Agent Interaction* (pp. 71–74). https://doi.org/10.1145/2974804.2980491

Jackson, L. A., & Wang, J.-L. (2013). Cultural differences in social networking site use: A comparative study of China and the United States. *Computers in Human Behavior, 29*(3), 910–921. https://doi.org/10.1016/j.chb.2012.11.024

Jagne, J., & Smith-Atakan, A. S. G. (2006). Cross-cultural interface design strategy. *Universal Access in the Information Society, 5*(3), 299–305. https://doi.org/10.1007/s10209-006-0048-6

Jamil, I., Montero, C. S., Perry, M., O'Hara, K., Karnik, A., Pihlainen, K., Marshall, M. T., Jha, S., Gupta, S., & Subramanian, S. (2017). Collaborating around digital tabletops: Children's physical strategies from India, the UK and Finland. *ACM Transactions on Computer-Human Interaction, 24*(3), 1–30. https://doi.org/10.1145/3058551

Javidan, M., House, R. J., Dorfman, P. W., Hanges, P. J., & Sully de Luque, M. (2006). Conceptualizing and measuring cultures and their consequences: A comparative review of GLOBE's and Hofstede's approaches. *Journal of International Business Studies, 37*(6), 897–914. https://doi.org/10.1057/palgrave.jibs.8400234

Jhangiani, I., & Smith-Jackson, D. T. (2007). *Comparison of mobile phone user interface design preferences: Perspectives from nationality and disability culture.*

Jones, M. L. (2007). *Hofstede - Culturally questionable?* Oxford Business & Economics Conference.

Kaarst-Brown, M. L., & Guzman, I. R. (2014). Cultural richness versus cultural large scale insights: Culture, globalization, and it workers. In *Proceedings of the 52nd ACM Conference on Computers and People Research* (pp. 55–58). https://doi.org/10.1145/2599990.2600002

Kahn, Z., & Burrell, J. (2021). A sociocultural explanation of internet-enabled work in rural regions. *ACM Transactions on Computer-Human Interaction, 28*(3), 1–22. https://doi.org/10.1145/344 3705

Kamppuri, M. (2012). Because deep down, we are not the same: Values in cross-cultural design. *Interactions, 19*(2), 65–68. https://doi.org/10.1145/2090150.2090166

Kamppuri, M., Bednarik, R., & Tukiainen, M. (2006). The expanding focus of HCI: Case culture. In *Proceedings of the 4th Nordic Conference on Human-Computer Interaction: Changing Roles* (pp. 405–408). https://doi.org/10.1145/1182475.1182523

Karran, A. J., Fairclough, S. H., & Gilleade, K. (2015). A framework for psychophysiological classification within a cultural heritage context using interest. *ACM Transactions on Computer-Human Interaction, 21*(6), 1–19. https://doi.org/10.1145/2687925

Katz, S. (2009). *Cultural aging. Life course, lifestyles, and senior worlds.* University of Toronto Press.

Kawu, A. A., Orji, R., Awal, A., & Gana, U. (2018). Personality, culture and password behavior: A relationship study. In *Proceedings of the Second African Conference for Human Computer Interaction: Thriving Communities* (pp. 1–4). https://doi.org/10.1145/3283458.3283530

Kayan, S., Fussell, S. R., & Setlock, L. D. (2006). Cultural differences in the use of instant messaging in Asia and North America. In *Proceedings of the 2006 20th Anniversary Conference on Computer Supported Cooperative Work* (pp. 525–528). https://doi.org/10.1145/1180875.1180956

Kayes, I., Kourtellis, N., Quercia, D., Iamnitchi, A., & Bonchi, F. (2015). Cultures in community question answering. In *Proceedings of the 26th ACM Conference on Hypertext & Social Media - HT'15* (pp. 175–184). https://doi.org/10.1145/2700171.2791034

Kefalidou, G., Hoang, V., Padiyil, M. H., & Bedwell, B. (2015). Retroflection: Self-reflection for knowledge and culture sustainability. In *Proceedings of the 17th International Conference on Human-Computer Interaction with Mobile Devices and Services Adjunct* (pp. 1046–1049). https://doi.org/10.1145/2786567.2794329

Keith, K. (Ed.). (2019). *Cross-cultural psychology. Contemporary themes and perspectives.* Wiley, Blackwell.

Kerne, A., Hamilton, W. A., & Toups, Z. O. (2012). Culturally based design: Embodying transsurface interaction in rummy. In *Proceedings of the ACM 2012 Conference on Computer Supported Cooperative Work* (pp. 509–518). https://doi.org/10.1145/2145204.2145284

Keyes, O., Peil, B., Williams, R. M., & Spiel, K. (2020). Reimagining (women's) health: HCI, gender and essentialised embodiment. *ACM Transactions on Computer-Human Interaction, 27*(4), 1–42. https://doi.org/10.1145/3404218

Khaled, R. (2008). *Culturally-relevant persuasive technology.* Victoria University of Wellington.

Khaled, R., Barr, P., Fischer, R., Noble, J., & Biddle, R. (2006a). Factoring culture into the design of a persuasive game. In *Proceedings of the 20th Conference of the Computer-Human Interaction Special Interest Group (CHISIG) of Australia on Computer-Human Interaction: Design: Activities, Artefacts and Environments - OZCHI'06* (p. 213). https://doi.org/10.1145/1228175.1228213

Khaled, R., Biddle, R., Noble, J., Barr, P., & Fischer, R. (2006b). Persuasive interaction for collectivist cultures. In *Conferences in Research and Practice in Information Technology Series* (pp. 63–70).

Khan, T., Pitts, M., & Williams, M. A. (2016). Cross-cultural differences in automotive HMI design: A comparative study between UK and Indian users' design preferences. *The Journal of User Experience, 11*(2).

Khooshabeh, P., Dehghani, M., Nazarian, A., & Gratch, J. (2017). The cultural influence model: When accented natural language spoken by virtual characters matters. *AI & Society, 32*(1), 9–16. https://doi.org/10.1007/s00146-014-0568-1

Khurana, M., Chen, Z., Byrne, D., & Bai, Y. (2021). SneezeLove: Embodying cultural superstitions in connected devices. In *Designing Interactive Systems Conference 2021* (pp. 1082–1086). https://doi.org/10.1145/3461778.3462118

Kim, H. S. (2002). We talk, therefore we think? A cultural analysis of the effect of talking on thinking. *Journal of Personality and Social Psychology, 83*(4), 828–842. https://doi.org/10.1037/0022-3514.83.4.828

Kim, H. S., & Sasaki, J. Y. (2014). Cultural neuroscience: Biology of the mind in cultural contexts. *Annual Review of Psychology, 65*(1), 487–514. https://doi.org/10.1146/annurev-psych-010213-115040

Kim, T., Lee, D., Hyun, S. J., & Doh, Y. Y. (2019). UrbanSocialRadar: A place-aware social matching model for estimating serendipitous interaction willingness in Korean cultural context. *International Journal of Human-Computer Studies, 125*, 81–103. https://doi.org/10.1016/j.ijhcs.2018.12.011

Kimura-Thollander, P., & Kumar, N. (2019). Examining the "Global" language of Emojis: Designing for cultural representation. In *Proceedings of the 2019 CHI Conference on Human Factors in Computing Systems* (pp. 1–14). https://doi.org/10.1145/3290605.3300725

Kimura, H., & Nakajima, T. (2010). EcoIsland: A persuasive application to motivate sustainable behavior in collectivist cultures. In *Proceedings of the 6th Nordic Conference on Human-Computer Interaction: Extending Boundaries* (pp. 703–706). https://doi.org/10.1145/1868914.1869009

Kistler, F., Endrass, B., Damian, I., Dang, C. T., & André, E. (2012). Natural interaction with culturally adaptive virtual characters. *Journal on Multimodal User Interfaces, 6*(1–2), 39–47. https://doi.org/10.1007/s12193-011-0087-z

Kitayama, S., & Cohen, D. (Eds.). (2007). *Handbook of cultural psychology.* Guilford Press.

Kitayama, S., Duffy, S., Kawamura, T., & Larsen, J. T. (2003). Perceiving an object and its context in different cultures: A cultural look at new look. *Psychological Science, 14*(3), 201–206. https://doi.org/10.1111/1467-9280.02432

Kizilcec, R. F., & Cohen, G. L. (2017). Eight-minute self-regulation intervention raises educational attainment at scale in individualist but not collectivist cultures. *Proceedings of the National Academy of Sciences, 114*(17), 4348–4353. https://doi.org/10.1073/pnas.1611898114

Kleinsmith, A., De Silva, P. R., & Bianchi-Berthouze, N. (2006). Cross-cultural differences in recognizing affect from body posture. *Interacting with Computers, 18*(6), 1371–1389. https://doi.org/10.1016/j.intcom.2006.04.003

Kling, C. C., & Gottron, T. (n.d.-a). *Detecting culture in coordinates: Cultural areas in social media.*

Koay, K. Y., Sandhu, M. S., Tjiptono, F., & Watabe, M. (2022). Understanding employees' knowledge hiding behaviour: The moderating role of market culture. *Behaviour & Information Technology, 41*(4), 694–711. https://doi.org/10.1080/0144929X.2020.1831073

Koda, T., & Takeda, Y. (2018). Perception of culture-specific gaze behaviors of agents and gender effects. In *Proceedings of the 6th International Conference on Human-Agent Interaction* (pp. 138–143). https://doi.org/10.1145/3284432.3284472

Koda, T., Ishida, T., Rehm, M., & André, E. (2009). Avatar culture: Cross-cultural evaluations of avatar facial expressions. *AI & Society, 24*(3), 237–250. https://doi.org/10.1007/s00146-009-0214-5

Koda, T., Rehm, M., & André, E. (2008). Cross-cultural evaluations of avatar facial expressions designed by western designers. In H. Prendinger, J. Lester, & M. Ishizuka (Eds.), *Intelligent virtual agents* (Vol. 5208, pp. 245–252). Springer. https://doi.org/10.1007/978-3-540-85483-8_25

Komlodi, A., Hou, W., Preece, J., Druin, A., Golub, E., Alburo, J., Liao, S., Elkiss, A., & Resnik, P. (2007). Evaluating a cross-cultural children's online book community: Lessons learned for sociability, usability, and cultural exchange. *Interacting with Computers, 19*(4), 494–511. https://doi.org/10.1016/j.intcom.2007.03.001

Konstantakis, M., & Caridakis, G. (2020). Adding culture to UX: UX research methodologies and applications in cultural heritage. *Journal on Computing and Cultural Heritage, 13*(1), 1–17. https://doi.org/10.1145/3354002

Kontiza, K., Liapis, A., & Jones, C. E. (2020). Reliving the experience of visiting a gallery: Methods for evaluating informal learning in games for cultural heritage. In *International Conference on the Foundations of Digital Games* (pp. 1–11). https://doi.org/10.1145/3402942.3403009

Korte, J., Potter, L. E., & Nielsen, S. (2017). The impacts of deaf culture on designing with deaf children. In *Proceedings of the 29th Australian Conference on Computer-Human Interaction* (pp. 135–142). https://doi.org/10.1145/3152771.3152786

Krekhov, A., Emmerich, K., Fuchs, J., & Krueger, J. H. (2022). Interpolating happiness: Understanding the intensity gradations of face Emojis across cultures. In *CHI Conference on Human Factors in Computing Systems* (pp. 1–17). https://doi.org/10.1145/3491102.3517661

Kroeber, A., & Kluckhohn, C. (1985). *Culture: A critical review of concepts and definitions*. Vintage Books.

Kuhn, T. (1996). *The structure of scientific revolutions*. The University of Chicago Press.

Kyriakoullis, L., & Zaphiris, P. (2016). Culture and HCI: A review of recent cultural studies in HCI and social networks. *Universal Access in the Information Society, 15*(4), 629–642. https://doi.org/10.1007/s10209-015-0445-9

Lachner, F., Nguyen, M.-A., & Butz, A. (2018). Culturally sensitive user interface design: A case study with German and Vietnamese users. In *Proceedings of the Second African Conference for Human Computer Interaction: Thriving Communities* (pp. 1–12). https://doi.org/10.1145/3283458.3283459

Ladd, P. (2003). *Understanding deaf culture*. Multilingual Matters.

Latimer, J., & Miele, M. (2013). Naturecultures? Science, affect and the non-human. *Theory, Culture & Society, 30*(7–8), 5–31. https://doi.org/10.1177/0263276413502088

Leal, D. D. C., Krüger, M., Teles, V. T. E., Teles, C. A. T. E., Cardoso, D. M., Randall, D., & Wulf, V. (2021). Digital technology at the edge of capitalism: Experiences from the Brazilian Amazon Rainforest. *ACM Transactions on Computer-Human Interaction, 28*(3), 1–39. https://doi.org/10.1145/3448072

Lee, H. R., & Sabanović, S. (2014). Culturally variable preferences for robot design and use in South Korea, Turkey, and the United States. In *Proceedings of the 2014 ACM/IEEE International Conference on Human-Robot Interaction* (pp. 17–24). https://doi.org/10.1145/2559636.2559676

Lee, H. R., Sung, J., Sabanovic, S., & Han, J. (2012). Cultural design of domestic robots: A study of user expectations in Korea and the United States. In *2012 IEEE RO-MAN: The 21st IEEE International Symposium on Robot and Human Interactive Communication* (pp. 803–808). https://doi.org/10.1109/ROMAN.2012.6343850

Lee, M. K., & Rich, K. (2021). Who is included in human perceptions of AI?: Trust and perceived fairness around healthcare AI and cultural mistrust. In *Proceedings of the 2021 CHI Conference on Human Factors in Computing Systems* (pp. 1–14). https://doi.org/10.1145/3411764.3445570

Lee, S. C., Stojmenova, K., Sodnik, J., Schroeter, R., Shin, J., & Jeon, M. (2019). Localization vs. internationalization: Research and practice on autonomous vehicles across different cultures. In *Proceedings of the 11th International Conference on Automotive User Interfaces and Interactive Vehicular Applications: Adjunct Proceedings* (pp. 7–12). https://doi.org/10.1145/3349263.3350760

Lee, S. H., Samdanis, M., & Gkiousou, S. (2014). Hybridizing food cultures in computer-mediated environments: Creativity and improvisation in Greek food blogs. *International Journal of Human-Computer Studies, 72*(2), 224–238. https://doi.org/10.1016/j.ijhcs.2013.08.007

Leidner & Kayworth. (2006). Review: A review of culture in information systems research: Toward a theory of information technology culture conflict. *MIS Quarterly, 30*(2), 357. https://doi.org/10.2307/25148735

Leiva, L. A., & Alabau, V. (2015). Automatic internationalization for just in time localization of web-based user interfaces. *ACM Transactions on Computer-Human Interaction, 22*(3), 1–32. https://doi.org/10.1145/2701422

Levine, R. (2006). *A geography of time: The temporal misadventures of a social psychologist, or how every culture keeps time just a little bit differently.* Oneworld.

Lewthwaite, S., & Sloan, D. (2016). Exploring pedagogical culture for accessibility education in computing science. In *Proceedings of the 13th International Web for All Conference* (pp. 1–4). https://doi.org/10.1145/2899475.2899490

Liao, H., Proctor, R. W., & Salvendy, G. (2008). Content preparation for cross-cultural e-commerce: A review and a model. *Behaviour & Information Technology, 27*(1), 43–61. https://doi.org/10.1080/01449290601088424

Lim, M. Y., Leichtenstern, K., Kriegel, M., Enz, S., Aylett, R., Vannini, N., Hall, L., & Rizzo, P. (2011). Technology-enhanced role-play for social and emotional learning context – Intercultural empathy. *Entertainment Computing, 2*(4), 223–231. https://doi.org/10.1016/j.entcom.2011.02.004

Lin, G. E., Mynatt, E. D., & Kumar, N. (2022). Investigating culturally responsive design for menstrual tracking and sharing practices among individuals with minimal sexual education. In *CHI Conference on Human Factors in Computing Systems* (pp. 1–15). https://doi.org/10.1145/3491102.3501824

Lin, H.-C., & Ho, W.-H. (2018). Cultural effects on use of online social media for health-related information acquisition and sharing in Taiwan. *International Journal of Human-Computer Interaction, 34*(11), 1063–1076. https://doi.org/10.1080/10447318.2017.1413790

Lindgren, A., Chen, F., Jordan, P. W., & Zhang, H. (2008). *Requirements for the design of advanced driver assistance systems – the differences between Swedish and Chinese drivers.*

Lindholm, C. (2008). *Culture and authenticity.* Blackwell Pub.

Lindtner, S., Anderson, K., & Dourish, P. (2012). Cultural appropriation: Information technologies as sites of transnational imagination. In *Proceedings of the ACM 2012 Conference on Computer Supported Cooperative Work* (pp. 77–86). https://doi.org/10.1145/2145204.2145220

Lindtner, S., Nardi, B., Wang, Y., Mainwaring, S., Jing, H., & Liang, W. (2008). A hybrid cultural ecology: World of warcraft in China. In *Proceedings of the 2008 ACM Conference on Computer Supported Cooperative Work* (pp. 371–382). https://doi.org/10.1145/1460563.1460624

Linxen, S., Cassau, V., & Sturm, C. (2021). Culture and HCI: A still slowly growing field of research. Findings from a systematic, comparative mapping review. In *Proceedings of the XXI International Conference on Human Computer Interaction* (pp. 1–5). https://doi.org/10.1145/3471391.3471421

Lionelle, A., Grinslad, J., & Beveridge, J. R. (2020). CS 0: Culture and coding. In *Proceedings of the 51st ACM Technical Symposium on Computer Science Education* (pp. 227–233). https://doi.org/10.1145/3328778.3366795

Liu, J., Rau, P.-L.P., & Wendler, N. (2015). Trust and online information-sharing in close relationships: A cross-cultural perspective. *Behaviour & Information Technology, 34*(4), 363–374. https://doi.org/10.1080/0144929X.2014.937458

Liu, S.-Y. (Cyn), Bardzell, J., & Bardzell, S. (2018). Photography as a design research tool into nature culture. In *Proceedings of the 2018 Designing Interactive Systems Conference* (pp. 777–789). https://doi.org/10.1145/3196709.3196819

Lofstrom, A. (2011). Culture as similarities and dynamics: Suggesting reinterpretations of Geert Hofstede's work. In *2011 Second International Conference on Culture and Computing* (pp. 133–134). https://doi.org/10.1109/Culture-Computing.2011.36

Lottridge, D., Chignell, M., & Yasumura, M. (2012). Identifying emotion through implicit and explicit measures: Cultural differences, cognitive load, and immersion. *IEEE Transactions on Affective Computing, 3*(2), 199–210. https://doi.org/10.1109/T-AFFC.2011.36

Lu, F., Tian, F., Jiang, Y., Cao, X., Luo, W., Li, G., Zhang, X., Dai, G., & Wang, H. (2011). Shadow-Story: Creative and collaborative digital storytelling inspired by cultural heritage. In *Proceedings of the SIGCHI Conference on Human Factors in Computing Systems* (pp. 1919–1928). https://doi.org/10.1145/1978942.1979221

Lu, X., Chen, Y., & Epstein, D. A. (2021). How cultural norms influence persuasive design: A study on Chinese food journaling apps. In *Designing Interactive Systems Conference 2021* (pp. 619–637). https://doi.org/10.1145/3461778.3462142

Lu, Z., Annett, M., Fan, M., & Wigdor, D. (2019). "I feel it is my responsibility to stream": Streaming and engaging with intangible cultural heritage through livestreaming. In *Proceedings of the 2019 CHI Conference on Human Factors in Computing Systems* (pp. 1–14). https://doi.org/10.1145/3290605.3300459

Lucas, G. M., Boberg, J., Traum, D., Artstein, R., Gratch, J., Gainer, A., Johnson, E., Leuski, A., & Nakano, M. (2018). Culture, errors, and rapport-building dialogue in social agents. In *Proceedings of the 18th International Conference on Intelligent Virtual Agents* (pp. 51–58). https://doi.org/10.1145/3267851.3267887

Lyu, Y., & Carroll, J. M. (2022). Cultural influences on Chinese Citizens' adoption of digital contact tracing: A human infrastructure perspective. In *CHI Conference on Human Factors in Computing Systems* (pp. 1–17). https://doi.org/10.1145/3491102.3517572

MacDonald, M. C., St-Cyr, O., Gray, C. M., Potter, L. E., Lallemand, C., Vasilchenko, A., Sin, J., Carter, A. R. L., Pitt, C., Sari, E., Ranjan Padhi, D., & Pillai, A. G. (2022). EduCHI 2022: 4th annual symposium on HCI education. In *CHI Conference on Human Factors in Computing Systems Extended Abstracts* (pp. 1–5). https://doi.org/10.1145/3491101.3503703

MacDorman, K. F., Vasudevan, S. K., & Ho, C.-C. (2009). Does Japan really have robot mania? Comparing attitudes by implicit and explicit measures. *AI & Society, 23*(4), 485–510. https://doi.org/10.1007/s00146-008-0181-2

Mankoff, J., Hayes, G. R., & Kasnitz, D. (2010). Disability studies as a source of critical inquiry for the field of assistive technology. In *Proceedings of the 12th International ACM SIGAC-CESS Conference on Computers and Accessibility* (pp. 3–10). https://doi.org/10.1145/1878803.1878807

Manohar, A. K. (2018). Story culture framework: A cross cultural study. In *Proceedings of the 9th Indian Conference on Human-Computer Interaction* (pp. 44–52). https://doi.org/10.1145/3297121.3297122

Manovich, L. (2016). The science of culture? Social computing, digital humanities and cultural analytics. *Journal of Cultural Analytics.* https://doi.org/10.22148/16.004

Marcos, M.-C., Garcia-Gavilanes, R., Bataineh, E., & Pasarin, L. (n.d.). *Using eye tracking to identify cultural differences in information seeking behavior.*

Marcus, A. (2010). *Cross-cultural user-interface design for work, home, play, and on the way.*

Marcus, A. (n.d.). *Cross-cultural user-experience design: What? So What? Now What?*

Marcus, A., & Baumgartner, V.-J. (2004). A practical set of culture dimensions for global user-interface development. In M. Masoodian, S. Jones, & B. Rogers (Eds.), *Computer human interaction* (Vol. 3101, pp. 252–261). Springer. https://doi.org/10.1007/978-3-540-27795-8_26

Marcus, A., & Gould, E. W. (2000). Cultural dimensions and global web user-interface design. *Interactions*, 32–46.

Marcus, A., Kurosu, M., Ma, X., & Hashizume, A. (2017). Cuteness engineering. *Springer International Publishing.* https://doi.org/10.1007/978-3-319-61961-3

Markussen, T., & Krogh, P. G. (2008). *Mapping cultural frame shifting in interaction design with blending theory.*

Marsden, G., Maunder, A., & Parker, M. (2008). People are people, but technology is not technology. *Philosophical Transactions of the Royal Society A: Mathematical, Physical and Engineering Sciences, 366*(1881), 3795–3804. https://doi.org/10.1098/rsta.2008.0119

Martinviita, A., Kuure, L., & Luoma, P. (2015). Do we speak the same language?: Design goals and culture clashes in an online forum for young people. In *Proceedings of the 7th International Conference on Communities and Technologies* (pp. 69–78). https://doi.org/10.1145/2768545.2768550

Marwick, A. (1999).*The sixties: Social and cultural transformation in Britain, France, Italy and the United States, 1958–74.* Bloomsbury Reader.

Mascarenhas, S., Degens, N., Paiva, A., Prada, R., Hofstede, G. J., Beulens, A., & Aylett, R. (2016). Modeling culture in intelligent virtual agents: From theory to implementation. *Autonomous Agents and Multi-Agent Systems, 30*(5), 931–962. https://doi.org/10.1007/s10458-015-9312-6

Massey, A. P., Hung, Y.-T. C., Montoya-Weiss, M., & Ramesh, V. (n.d.). *When culture and style aren't about clothes: Perceptions of task-technology "Fit" in global virtual teams.*

Maye, L. A., Bouchard, D., Avram, G., & Ciolfi, L. (2017). Supporting cultural heritage professionals adopting and shaping interactive technologies in museums. In *Proceedings of the 2017 Conference on Designing Interactive Systems* (pp. 221–232). https://doi.org/10.1145/3064663.3064753

McAlear, F., Her Many Horses, I., Casao, M., & Luebker, R. (2022). Code red: Culturally revitalizing computing courses in native American-serving schools. In *Proceedings of the 53rd ACM Technical Symposium on Computer Science Education* (Vol. 2, pp. 1057–1058). https://doi.org/10.1145/3478432.3499232

McArthur, J. A. (2009). Digital subculture: A Geek meaning of style. *Journal of Communication Inquiry, 33*(1), 58–70. https://doi.org/10.1177/0196859908325676

McGookin, D., Tahiroğlu, K., Vaittinen, T., Kytö, M., Monastero, B., & Carlos Vasquez, J. (2019). Investigating tangential access for location-based digital cultural heritage applications. *International Journal of Human-Computer Studies, 122*, 196–210. https://doi.org/10.1016/j.ijhcs.2018.09.009

McKenna, P. E., Ghosh, A., Aylett, R., Broz, F., & Rajendran, G. (2018). Cultural social signal interplay with an expressive robot. In *Proceedings of the 18th International Conference on Intelligent Virtual Agents* (pp. 211–218). https://doi.org/10.1145/3267851.3267905

McSweeney, B. (n.d.). Hofstede's model of national cultural differences and their consequences: A triumph of faith—a failure of analysis. *Human Relations.*

Meriläinen, M. (2023). Young people's engagement with digital gaming cultures – validating and developing the digital gaming relationship theory. *Entertainment Computing, 44*, 100538. https://doi.org/10.1016/j.entcom.2022.100538

Merritt, S., & Stolterman, E. (2012). Cultural hybridity in participatory design. In *Proceedings of the 12th Participatory Design Conference: Exploratory Papers, Workshop Descriptions, Industry Cases* (Vol. 2, pp. 73–76). https://doi.org/10.1145/2348144.2348168

Metcalf, H. E., Crenshaw, T. L., Chambers, E. W., & Heeren, C. (2018). Diversity across a decade: A case study on undergraduate computing culture at the University of Illinois. In *Proceedings of the 49th ACM Technical Symposium on Computer Science Education* (pp. 610–615). https://doi.org/10.1145/3159450.3159497

Meyers, E. M., Nathan, L. P., & Tulloch, B. (2019). Designing picturebook apps: Valuing culture & community. In *Proceedings of the 9th International Conference on Communities & Technologies - Transforming Communities* (pp. 14–23). https://doi.org/10.1145/3328320.3328377

Michailidou, E., Parmaxi, A., & Zaphiris, P. (2015). Culture effects in online social support for older people: Perceptions and experience. *Universal Access in the Information Society, 14*(2), 281–293. https://doi.org/10.1007/s10209-014-0346-3

Miller, D. (Ed.). (2001). *Car cultures*. Berg.

Moore, J. D. (2009). *Visions of culture: An introduction to anthropological theories and theorists* (3rd ed). AltaMira Press.

Mora-Cantallops, M., Muñoz, E., Santamaría, R., & Sánchez-Alonso, S. (2021). Identifying communities and fan practices in online retrogaming forums. *Entertainment Computing, 38*, 100410. https://doi.org/10.1016/j.entcom.2021.100410

Morozov, E. (2013). *To save everything, click here. The folly of technological solutionism*. Public Affairs.

Muratovski, G., & Vogel, C. (Eds.). (2019). *Design discourse on culture and society: Re:Research* (Vol. 5). Intellect Books. https://doi.org/10.2307/j.ctv36xw0zv

Nadkarni, A., & Hofmann, S. G. (2012). Why do people use Facebook? *Personality and Individual Differences, 52*(3), 243–249. https://doi.org/10.1016/j.paid.2011.11.007

Namara, M., Wilkinson, D., Lowens, B. M., Knijnenburg, B. P., Orji, R., & Sekou, R. L. (2018). Cross-cultural perspectives on eHealth privacy in Africa. In *Proceedings of the Second African Conference for Human Computer Interaction: Thriving Communities* (pp. 1–11). https://doi.org/10.1145/3283458.3283472

Naneva, S., Sarda Gou, M., Webb, T. L., & Prescott, T. J. (2020). A systematic review of attitudes, anxiety, acceptance, and trust towards social robots. *International Journal of Social Robotics, 12*(6), 1179–1201. https://doi.org/10.1007/s12369-020-00659-4

Ng, K.-Y., Van Dyne, L., & Ang, S. (2012). Cultural intelligence: A review, reflections, and recommendations for future research. In A. M. Ryan, F. T. L. Leong, & F. L. Oswald (Eds.), *Conducting multinational research: Applying organizational psychology in the workplace* (pp. 29–58). American Psychological Association. https://doi.org/10.1037/13743-002

Nguyen, D. T., & Fussell, S. R. (2012). How did you feel during our conversation?: Retrospective analysis of intercultural and same-culture instant messaging conversations. In *Proceedings of the ACM 2012 Conference on Computer Supported Cooperative Work* (pp. 117–126). https://doi.org/10.1145/2145204.2145225

Nielsen, J., Del Galdo, E. M., & Sprung, R. C. (1990). Designing for international use. In *Conference on Human Factors in Computing Systems - Proceedings*, April, 291–294. https://doi.org/10.1145/97243.97298

Nisbett, R. E. (2003). *The geography of thought: How Asians and Westerners think differently– and why*. Free Press.

Nisbett, R. E., Peng, K., Choi, I., & Norenzayan, A. (2001). Culture and systems of thought: Holistic versus analytic cognition. *Psychological Review, 108*(2), 291–310. https://doi.org/10.1037/0033-295X.108.2.291

Noiwan, J., & Norcio, A. F. (2006). Cultural differences on attention and perceived usability: Investigating color combinations of animated graphics. *International Journal of Human-Computer Studies, 64*(2), 103–122. https://doi.org/10.1016/j.ijhcs.2005.06.004

Noorbehbahani, F., & Salehi, F. (2021). A serious game to extract Hofstede's cultural dimensions at the individual level. *User Modeling and User-Adapted Interaction, 31*(2), 225–259. https://doi.org/10.1007/s11257-020-09280-6

Norman, D. (1998). *The invisible computer.* The MIT Press.

Norman, D. (2004). *Emotional design.* Basic Books.

Not, E., & Petrelli, D. (2019). Empowering cultural heritage professionals with tools for authoring and deploying personalised visitor experiences. *User Modeling and User-Adapted Interaction, 29*(1), 67–120. https://doi.org/10.1007/s11257-019-09224-9

Ntakolia, C., Dimas, G., & Iakovidis, D. K. (2022). User-centered system design for assisted navigation of visually impaired individuals in outdoor cultural environments. *Universal Access in the Information Society, 21*(1), 249–274. https://doi.org/10.1007/s10209-020-00764-1

O'Kane, A. A., Aliomar, A., Zheng, R., Schulte, B., & Trombetta, G. (2019). Social, cultural and systematic frustrations motivating the formation of a DIY hearing loss hacking community. In *Proceedings of the 2019 CHI Conference on Human Factors in Computing Systems* (pp. 1–14). https://doi.org/10.1145/3290605.3300531

O'Leary, T. K., Stowell, E., Kimani, E., Parmar, D., Olafsson, S., Hoffman, J., Parker, A. G., Paasche-Orlow, M. K., & Bickmore, T. (2020). Community-based cultural tailoring of virtual agents. In *Proceedings of the 20th ACM International Conference on Intelligent Virtual Agents* (pp. 1–8). https://doi.org/10.1145/3383652.3423875

Obermeyer, Z., Powers, B., Vogeli, C., & Mullainathan, S. (2019). Dissecting racial bias in an algorithm used to manage the heath of populations. *Science, 366*, 447–453.

Oguamanam, V., Lee, T., McKlin, T., Cochran, Z., Abowd, G., & DiSalvo, B. (2020). Cultural clash: Exploring how studio-based pedagogy impacts learning for students in HCI classrooms. In *Proceedings of the 2020 ACM Designing Interactive Systems Conference* (pp. 1131–1142). https://doi.org/10.1145/3357236.3395544

Olasina, G., & Mutula, S. (2015). The influence of national culture on the performance expectancy of e-parliament adoption. *Behaviour & Information Technology, 34*(5), 492–505. https://doi.org/10.1080/0144929X.2014.1003326

Oliveira, N., Andrade, N., & Reinecke, K. (2016). Participation differences in Q&A sites across countries: Opportunities for cultural adaptation. In *Proceedings of the 9th Nordic Conference on Human-Computer Interaction* (pp. 1–10). https://doi.org/10.1145/2971485.2971520

Orji, R., & Mandryk, R. L. (2014). Developing culturally relevant design guidelines for encouraging healthy eating behavior. *International Journal of Human-Computer Studies, 72*(2), 207–223. https://doi.org/10.1016/j.ijhcs.2013.08.012

Ornelas, M. L., Smith, G. B., & Mansouri, M. (2022). Redefining culture in cultural robotics. *AI & Society.* https://doi.org/10.1007/s00146-022-01476-1

Oshlyansky, L., Thimbleby, H., & Cairns, P. (2004). Breaking affordance: Culture as context. In *Proceedings of the Third Nordic Conference on Human-Computer Interaction* (pp. 81–84). https://doi.org/10.1145/1028014.1028025

Oudshoorn, N., & Pinch, T. (Eds.). (2003). *How users matter.* The MIT Press. https://doi.org/10.1353/tech.2006.0041

Oyugi, C., Dunckley, L., & Smith, A. (n.d.). *Evaluation methods and cultural differences: Studies across three continents.*

Özkan, T., Lajunen, T., Chliaoutakis, J. E., Parker, D., & Summala, H. (2006). Cross-cultural differences in driving behaviours: A comparison of six countries. *Transportation Research Part F: Traffic Psychology and Behaviour, 9*(3), 227–242. https://doi.org/10.1016/j.trf.2006.01.002

Papangelis, K., Chamberlain, A., & Liang, H. N. (2016). New directions for preserving intangible cultural heritage through the use of mobile technologies. In *Proceedings of the 18th International Conference on Human-Computer Interaction with Mobile Devices and Services Adjunct* (pp. 964–967). https://doi.org/10.1145/2957265.2962643

Pappachan, P., & Ziefle, M. (2008). Cultural influences on the comprehensibility of icons in mobile–computer interaction. *Behaviour & Information Technology, 27*(4), 331–337. https://doi.org/10.1080/01449290802228399

Parker, A. G., & Grinter, R. E. (2014). Collectivistic health promotion tools: Accounting for the relationship between culture, food and nutrition. *International Journal of Human-Computer Studies, 72*(2), 185–206. https://doi.org/10.1016/j.ijhcs.2013.08.008

Parmaxi, A., & Zaphiris, P. (2016). Computer-mediated communication in computer-assisted language learning: Implications for culture-centered design. *Universal Access in the Information Society, 15*(1), 169–177. https://doi.org/10.1007/s10209-015-0405-4

Patel, N. J., Clawson, J., Kang, N., Choi, S., & Starner, T. (2010). A study of cultural effects on mobile-collocated group photo sharing. In *Proceedings of the 16th ACM International Conference on Supporting Group Work - GROUP'10* (p. 121). https://doi.org/10.1145/1880071.1880091

Paterson, B., Winschiers-Theophilus, H., Dunne, T. T., Schinzel, B., & Underhill, L. G. (2011). Interpretation of a cross-cultural usability evaluation: A case study based on a hypermedia system for rare species management in Namibia. *Interacting with Computers, 23*(3), 239–246. https://doi.org/10.1016/j.intcom.2011.03.002

Payne, W. C., Bergner, Y., West, M. E., Charp, C., Shapiro, R. B. B., Szafir, D. A., Taylor, E. V., & DesPortes, K. (2021). danceON: Culturally responsive creative computing. In *Proceedings of the 2021 CHI Conference on Human Factors in Computing Systems* (pp. 1–16). https://doi.org/10.1145/3411764.3445149

Pearson, J., Robinson, S., Reitmaier, T., Jones, M., & Joshi, A. (2019). Diversifying future-making through iterative design. *ACM Transactions on Computer-Human Interaction, 26*(5), 1–21. https://doi.org/10.1145/3341727

Pei, L., & Nardi, B. (2019). We did it right, but it was still wrong: Toward assets-based design. In *Extended Abstracts of the 2019 CHI Conference on Human Factors in Computing Systems* (pp. 1–11). https://doi.org/10.1145/3290607.3310434

Peltonen, E., Lagerspetz, E., Hamberg, J., Mehrotra, A., Musolesi, M., Nurmi, P., & Tarkoma, S. (2018). The hidden image of mobile apps: Geographic, demographic, and cultural factors in mobile usage. In *Proceedings of the 20th International Conference on Human-Computer Interaction with Mobile Devices and Services* (pp. 1–12). https://doi.org/10.1145/3229434.3229474

Pereira, R., & Cecília Calani Baranauskas, M. (2015). A value-oriented and culturally informed approach to the design of interactive systems. *International Journal of Human-Computer Studies, 80*, 66–82. https://doi.org/10.1016/j.ijhcs.2015.04.001

Peters, A., Winschiers-Theophilus, H., & Mennecke, B. (2013). *Bridging the digital divide through Facebook friendships: A cross-cultural study.*

Peters, S. (2000a). Is there a disability culture? A syncretisation of three possible world views. *Disability & Society, 15*(4), 583–601. https://doi.org/10.1080/09687590050058198

Peters, S. (2000b). Is there a disability culture?: A syncretisation of three possible world views. *Overcoming Disabling Barriers: 18 Years of Disability and Society, 15*(4), 583–601. https://doi.org/10.4324/9780203965030

Petrelli, D. (2019). Tangible interaction meets material culture: Reflections on the meSch project. *Interactions, 26*(5), 34–39. https://doi.org/10.1145/3349268

Petrie, H., & Merdenyan, B. (2016). Cultural and gender differences in password behaviors: Evidence from China, Turkey and the UK. In *Proceedings of the 9th Nordic Conference on Human-Computer Interaction* (pp. 1–10). https://doi.org/10.1145/2971485.2971563

Petrie, H., Weber, G., Jadhav, C., & Darzentas, J. S. (2018). Issues of culture in designing for accessibility. In T. Clemmensen, V. Rajamanickam, P. Dannenmann, H. Petrie, & M. Winckler (Eds.),
Global thoughts, local designs (Vol. 10774, pp. 55–67). Springer International Publishing. https://
doi.org/10.1007/978-3-319-92081-8_6

Phillips, W. (2019) Cross-cultural differences in visual perception of color, illusions, depth, and
pictures. In K. Keith (Ed.), *Cross-cultural psychology. Contemporary themes and perspectives*
(pp. 287–309). Wiley, Blackwell.

Poon, S. S., Thomas, R. C., Aragon, C. R., & Lee, B. (n.d.). *Context-linked virtual assistants for
distributed teams: An astrophysics case study.*

Poulopoulos, V., & Wallace, M. (2022). Digital technologies and the role of data in cultural heritage:
The past, the present, and the future. *Big Data and Cognitive Computing, 6*(3), 73. https://doi.org/
10.3390/bdcc6030073

Prabhakar, A. S., Maris, E., & Medhi Thies, I. (2021). Toward understanding the cultural influences
on social media use of middle class mothers in India. In *Extended Abstracts of the 2021 CHI
Conference on Human Factors in Computing Systems* (pp. 1–7). https://doi.org/10.1145/3411763.
3451779

Prendinger, H., & Ishizuka, M. (Eds.). (2004). *Life-like characters: Tools, affective functions, and
applications.* Springer. https://doi.org/10.1007/978-3-662-08373-4

Proctor, R. W., Nof, S. Y., Yih, Y., Balasubramanian, P., Busemeyer, J. R., Carayon, P., Chiu, C.-
Y., Farahmand, F., Gonzalez, C., Gore, J., Landry, S. J., Lehto, M., Rau, P.-L., Rouse, W., Tay,
L., Vu, K.-P.L., Woo, S. E., & Salvendy, G. (2011). Understanding and improving cross-cultural
decision making in design and use of digital media: A research agenda. *International Journal of
Human-Computer Interaction, 27*(2), 151–190. https://doi.org/10.1080/10447318.2011.537175

Profita, H. P., Stangl, A., Matuszewska, L., Sky, S., & Kane, S. K. (2016). Nothing to hide: Aesthetic
customization of hearing aids and cochlear implants in an online community. In *Proceedings of
the 18th International ACM SIGACCESS Conference on Computers and Accessibility* (pp. 219–
227). https://doi.org/10.1145/2982142.2982159

Puddephatt, A., Shaffir, W., & Kleinknecht, S. (Eds.). (2009). *Ethnographies revisited. Constructing
theory in the field.* Routledge

Punuru, A., Cheng, T.-W., Ghosh, I., Page, X., & Mondal, M. (2020). Cultural norms and interpersonal relationships: Comparing disclosure behaviors on Twitter. In *Conference Companion
Publication of the 2020 on Computer Supported Cooperative Work and Social Computing*
(pp. 371–375). https://doi.org/10.1145/3406865.3418341

Pyae, A. (2018). Understanding the role of culture and cultural attributes in digital game localization.
Entertainment Computing, 26, 105–116. https://doi.org/10.1016/j.entcom.2018.02.004

Pyae, A., & Potter, L. E. (2017). Does culture matter?: Understanding the impact of cultural contents
in digital games on older people. In *Proceedings of the 29th Australian Conference on Computer-
Human Interaction* (pp. 607–611). https://doi.org/10.1145/3152771.3156181

Pyae, A., Zaw, H. H., & Khine, M. T. (2018). Understanding the impact of cultural contents in digital games on players' engagement, enjoyment, and motivation in gameplay. In *Proceedings of
the 30th Australian Conference on Computer-Human Interaction* (pp. 83–87). https://doi.org/10.
1145/3292147.3292192

Qin, X., Tan, C.-W., Law, E. L.-C., Bødker, M., Clemmensen, T., Qu, H., & Chen, D. (2018).
Deciphering the role of context in shaping mobile phone usage: Design recommendations for
context-aware mobile services from a cross-cultural perspective. In *Proceedings of the Sixth
International Symposium of Chinese CHI* (pp. 39–48). https://doi.org/10.1145/3202667.3202673

Quappe, S., & Cantatore, G. (2005). *What is cultural awareness, anyway? How do I build it?*

Quattrone, G., Mashhadi, A., & Capra, L. (2014). Mind the map: The impact of culture and economic affluence on crowd-mapping behaviours. In *Proceedings of the 17th ACM Conference on Computer Supported Cooperative Work & Social Computing* (pp. 934–944). https://doi.org/10.1145/2531602.2531713

Raessens, J. (2006). Playful identities, or the ludification of culture. *Games and Culture, 1*(1), 52–57. https://doi.org/10.1177/1555412005281779

Ranasinghe, C., Holländer, K., Currano, R., Sirkin, D., Moore, D., Schneegass, S., & Ju, W. (2020). Autonomous vehicle-pedestrian interaction across cultures: Towards designing better external human machine interfaces (eHMIs). In *Extended Abstracts of the 2020 CHI Conference on Human Factors in Computing Systems* (pp. 1–8). https://doi.org/10.1145/3334480.3382957

Rapport, N., & Overing, J. (n.d.). *Social and cultural anthropology: The key concepts.*

Raptis, G. E., Fidas, C. A., & Avouris, N. M. (2016). Do field dependence-independence differences of game players affect performance and behaviour in cultural heritage games? In *Proceedings of the 2016 Annual Symposium on Computer-Human Interaction in Play* (pp. 38–43). https://doi.org/10.1145/2967934.2968107

Ratner, C., & Hui, L. (2003). Theoretical and methodological problems in cross-cultural psychology. *Journal for the Theory of Social Behaviour, 33*(1), 67–94. https://doi.org/10.1111/1468-5914.00206

Rau, P.-L. P. (Ed.). (2018). *Cross-Cultural Design. Applications in Cultural Heritage, Creativity and Social Development: 10th International Conference, CCD 2018, Held as Part of HCI International 2018, Las Vegas, NV, USA, July 15–20, 2018, Proceedings, Part II* (Vol. 10912). Springer International Publishing. https://doi.org/10.1007/978-3-319-92252-2

Rau, P.-L. P. (Ed.). (2020a). *Cross-Cultural Design. Applications in Health, Learning, Communication, and Creativity: 12th International Conference, CCD 2020, Held as Part of the 22nd HCI International Conference, HCII 2020, Copenhagen, Denmark, July 19–24, 2020, Proceedings, Part II* (Vol. 12193). Springer International Publishing. https://doi.org/10.1007/978-3-030-49913-6

Rau, P.-L. P. (Ed.). (2020b). *Cross-Cultural Design. User Experience of Products, Services, and Intelligent Environments: 12th International Conference, CCD 2020, Held as Part of the 22nd HCI International Conference, HCII 2020, Copenhagen, Denmark, July 19–24, 2020, Proceedings, Part I* (Vol. 12192). Springer International Publishing. https://doi.org/10.1007/978-3-030-49788-0

Rau, P.-L.P., Gao, Q., & Liang, S.-F.M. (2008). Good computing systems for everyone – how on earth? Cultural aspects? *Behaviour & Information Technology, 27*(4), 287–292. https://doi.org/10.1080/01449290701761250

Rau, P.-L. P., Liu, J., Verhasselt, S., Kato, T., & Schlick, C. M. (2011). Different time management behaviors of Germans, Chinese and Japanese. In *Proceedings of the ACM 2011 Conference on Computer Supported Cooperative Work* (pp. 701–704). https://doi.org/10.1145/1958824.1958949

Rauterberg, M. (Ed.). (2020). *Culture and Computing: 8th International Conference, C&C 2020, Held as Part of the 22nd HCI International Conference, HCII 2020, Copenhagen, Denmark, July 19–24, 2020, Proceedings* (Vol. 12215). Springer International Publishing. https://doi.org/10.1007/978-3-030-50267-6

Recabarren, M., & Nussbaum, M. (2010). Exploring the feasibility of web form adaptation to users' cultural dimension scores. *User Modeling and User-Adapted Interaction, 20*(1), 87–108. https://doi.org/10.1007/s11257-010-9071-7

Rehm, M., André, E., Bee, N., Endrass, B., Wissner, M., Nakano, Y., Nishida, T., & Huang, H.-H. (n.d.). *The CUBE-G approach – Coaching culture-specific nonverbal behavior by virtual agents*.

Reinecke, K., & Bernstein, A. (2011). Improving performance, perceived usability, and aesthetics with culturally adaptive user interfaces. *ACM Transactions on Computer-Human Interaction, 18*(2), 1–29. https://doi.org/10.1145/1970378.1970382

Reinecke, K., Nguyen, M. K., Bernstein, A., Näf, M., & Gajos, K. Z. (2013). Doodle around the world: Online scheduling behavior reflects cultural differences in time perception and group decision-making. In *Proceedings of the 2013 Conference on Computer Supported Cooperative Work* (pp. 45–54). https://doi.org/10.1145/2441776.2441784

Riddell, S., & Watson, N. (Eds.). (2014). *Disability*. Routledge.

Robinson, S., Bidwell, N. J., Cibin, R., Linehan, C., Maye, L., Mccarthy, J., Pantidi, N., & Teli, M. (2021). Rural Islandness as a Lens for (Rural) HCI. *ACM Transactions on Computer-Human Interaction, 28*(3), 1–32. https://doi.org/10.1145/3443704

Rodil, K., Rehm, M., & Winschiers-Theophilus, H. (2013). Homestead creator: Using card sorting in search for culture-aware categorizations of interface objects. In P. Kotzé, G. Marsden, G. Lindgaard, J. Wesson, & M. Winckler (Eds.), *Human-computer interaction – INTERACT 2013* (Vol. 8117, pp. 437–444). Springer. https://doi.org/10.1007/978-3-642-40483-2_30

Rodríguez, I., Puig, A., Tellols, D., & Samsó, K. (2020). Evaluating the effect of gamification on the deployment of digital cultural probes for children. *International Journal of Human-Computer Studies, 137*, 102395. https://doi.org/10.1016/j.ijhcs.2020.102395

Rogoff, B. (2003). *The cultural nature of human development*. Oxford University Press.

Rokkan, S. (1993). Interkulturelle, intersoziale und internationale Forschung cross-cultural, cross-societal and cross-national research. *Historical Social Research, 18*(2), 1. https://doi.org/10.12759/HSR.18.1993.2.6-54

Rose, H., & Yam, M. (2007). Sociocultural psychology: The dynamic interpendence among self systems and social systems. In S. Kitayama & D. Cohen (Eds.), *Handbook of cultural psychology* (pp. 3–40). The Guildford Press.

Ruiz-Calleja, A., Bote-Lorenzo, M. L., Asensio-Pérez, J. I., Villagrá-Sobrino, S. L., Alonso-Prieto, V., Gómez-Sánchez, E., García-Zarza, P., Serrano-Iglesias, S., & Vega-Gorgojo, G. (2023). Orchestrating ubiquitous learning situations about cultural heritage with casual learn mobile application. *International Journal of Human-Computer Studies, 170*, 102959. https://doi.org/10.1016/j.ijhcs.2022.102959

Russo, P., & Boor, S. (1993). How fluent is your interface?: Designing for international users. In *Proceedings of the SIGCHI Conference on Human Factors in Computing Systems - CHI'93* (pp. 342–347). https://doi.org/10.1145/169059.169274

Sá, G., Sylla, C., Martins, V., Caruso, A., & Menegazzi, D. (2019). Multiculturalism and creativity in storytelling—visual development of a digital manipulative for young children. In *Proceedings of the 2019 on Creativity and Cognition* (pp. 369–381). https://doi.org/10.1145/3325480.3326571

Sabie, D., Ekmekcioglu, C., & Ahmed, S. I. (2022). A decade of international migration research in HCI: Overview, challenges, ethics, impact, and future directions. *ACM Transactions on Computer-Human Interaction, 29*(4), 1–35. https://doi.org/10.1145/3490555

Sabie, D., Sabie, S., & Ahmed, S. I. (2020). Memory through design: Supporting cultural identity for immigrants through a paper-based home drafting tool. In *Proceedings of the 2020 CHI Conference on Human Factors in Computing Systems* (pp. 1–16). https://doi.org/10.1145/3313831.3376636

Sackmann, S. A. (2021). *Culture in organizations: Development*. Springer International Publishing. https://doi.org/10.1007/978-3-030-86080-6

Sahlins, M. D. (1976). *Culture and practical reason*. University of Chicago Press.

Salgado, L. C. de C., Leitão, C. F., & de Souza, C. S. (2013). *A journey through cultures: Metaphors for guiding the design of cross-cultural interactive systems.* Springer. https://doi.org/10.1007/978-1-4471-4114-3

Salgado, L., Pereira, R., & Gasparini, I. (2015). Cultural issues in HCI: Challenges and opportunities. In M. Kurosu (Ed.), *Human-computer interaction: Design and EValuation* (Vol. 9169, pp. 60–70). Springer International Publishing. https://doi.org/10.1007/978-3-319-20901-2_6

Samani, H., Saadatian, E., Pang, N., Polydorou, D., Fernando, O. N. N., Nakatsu, R., & Koh, J. T. K. V. (2013a). Cultural robotics: The culture of robotics and robotics in culture. *International Journal of Advanced Robotic Systems, 10*, 1–10. https://doi.org/10.5772/57260

Samani, H., Saadatian, E., Pang, N., Polydorou, D., Fernando, O. N. N., Nakatsu, R., & Koh, J. T. K. V. (2013b). Cultural robotics: The culture of robotics and robotics in culture. *International Journal of Advanced Robotic Systems, 10*(12), 400. https://doi.org/10.5772/57260

Samson, B. P. V., Shahid, S., Matsufuji, A., Wacharamanotham, C., Monserrat, T.-J. K. P., Sorathia, K., Ghazali, M., Miyafuji, S., Ahmed, N., Fujita, K., Islam, A. B. M. A. A., Sari, E., Mahmud, M., Tedjasaputra, A., Kim, J., Lee, U., Chintakovid, T., Wang, S.-M., Liu, Z., … Chen, B.-Y. (2020). Asian CHI symposium: HCI research from Asia and on Asian contexts and cultures. In *Extended Abstracts of the 2020 CHI Conference on Human Factors in Computing Systems* (pp. 1–6). https://doi.org/10.1145/3334480.3375068

Santos, E. E., Santos, E., Pan, L., Wilkinson, J. T., Thompson, J. E., & Korah, J. (2014). Infusing social networks with culture. *IEEE Transactions on Systems, Man, and Cybernetics: Systems, 44*(1), 1–17. https://doi.org/10.1109/TSMC.2013.2238922

Sarangapani, V., Kharrufa, A., Balaam, M., Leat, D., & Wright, P. (2016). Virtual.Cultural.Collaboration: Mobile phones, video technology, and cross-cultural learning. In *Proceedings of the 18th International Conference on Human-Computer Interaction with Mobile Devices and Services* (pp. 341–352). https://doi.org/10.1145/2935334.2935354

Sarangapani, V., Kharrufa, A., Leat, D., & Wright, P. (2019). Fostering deep learning in cross-cultural education through use of content-creation tools. In *Proceedings of the 10th Indian Conference on Human-Computer Interaction* (pp. 1–11). https://doi.org/10.1145/3364183.3364184

Satchell, C. (2008). Cultural theory and real world design—dystopian and utopian outcomes. *Culture and Technology.*

Sauer, J., Sonderegger, A., & Hoyos Álvarez, M. A. (2018). The influence of cultural background of test participants and test facilitators in online product evaluation. *International Journal of Human-Computer Studies, 111*, 92–100. https://doi.org/10.1016/j.ijhcs.2017.12.001

Sayago, S (Ed.) (2019). *Perspectives on human-computer interaction research with older people.* Springer Human-Computer Interaction Series.

Schein, E. (2004). *Organizational culture and leadership.* Wiley.

Schumacher, R. M. (Ed.). (2010). *The handbook of global user research.* Morgan Kaufmann.

Segal, T., & Vasilache, S. (2010). Plea against cultural stereotypes. In *Proceedings of the 3rd International Conference on Intercultural Collaboration* (pp. 275–278). https://doi.org/10.1145/1841853.1841906

Sengers, P. (2019). Practices for a machine culture: A case study of integrating cultural theory and artificial intelligence. *Surfaces, 8.* https://doi.org/10.7202/1065079ar

Setlock, L. D., & Fussell, S. R. (2010). What's it worth to you?: The costs and affordances of CMC tools to asian and american users. In *Proceedings of the 2010 ACM Conference on Computer Supported Cooperative Work* (pp. 341–350). https://doi.org/10.1145/1718918.1718979

Setlock, L. D., & Fussell, S. R. (2011a). Culture or fluency? Unpacking interactions between culture and communication medium. In *Conference on Human Factors in Computing Systems - Proceedings* (pp. 1137–1140). https://doi.org/10.1145/1978942.1979112

Setlock, L. D., Fussell, S. R., & Neuwirth, C. (2004). Taking it out of context: Collaborating within and across cultures in face-to-face settings and via instant messaging. In *Proceedings of the 2004 ACM Conference on Computer Supported Cooperative Work* (pp. 604–613). https://doi.org/10.1145/1031607.1031712

Setlock, L., & Fussell, S. (2011b). Culture or fluency?: Unpacking interactions between culture and communication medium. In *Proceedings of the SIGCHI Conference on Human Factors in Computing Systems* (pp. 1137–1140). https://doi.org/10.1145/1978942.1979112

Setlock, L., Quinones, P., & Fussell, S. (2007). Does culture interact with media richness? The effects of audio vs. video conferencing on Chinese and American Dyads. In *2007 40th Annual Hawaii International Conference on System Sciences (HICSS'07)* (pp. 13–13). https://doi.org/10.1109/HICSS.2007.182

Shah, H., Nersessian, N. J., Harrold, M. J., & Newstetter, W. (2012). Studying the influence of culture in global software engineering: Thinking in terms of cultural models. *Intercultural Communication*.

Shao, H. (2021). Do I prefer it?: The role of cultural heuristics in Chinese citizens' attitudes to COVID-19 rumors. In *Extended Abstracts of the 2021 CHI Conference on Human Factors in Computing Systems* (pp. 1–6). https://doi.org/10.1145/3411763.3451647

Sharp, H., Rogers, Y., & Preece, J. (2019). *Interaction design. Beyond human-computer interaction.* Wiley.

Sheldon, P., Herzfeldt, E., & Rauschnabel, P. A. (2020). Culture and social media: The relationship between cultural values and hashtagging styles. *Behaviour & Information Technology, 39*(7), 758–770. https://doi.org/10.1080/0144929X.2019.1611923

Shen, S.-T., Woolley, M., & Prior, S. (2006). Towards culture-centred design. *Interacting with Computers, 18*(4), 820–852. https://doi.org/10.1016/j.intcom.2005.11.014

Sheridan, T. B. (2020). A review of recent research in social robotics. *Current Opinion in Psychology, 36*, 7–12. https://doi.org/10.1016/j.copsyc.2020.01.003

Shi, Q. (2007). Cultural usability: The effects of culture on usability testing. In C. Baranauskas, P. Palanque, J. Abascal, & S. D. J. Barbosa (Eds.), *Human-computer interaction – INTERACT 2007* (Vol. 4663, pp. 611–616). Springer. https://doi.org/10.1007/978-3-540-74800-7_60

Shi, Q. (2008). A field study of the relationship and communication between Chinese evaluators and users in thinking aloud usability tests. In *Proceedings of the 5th Nordic Conference on Human-Computer Interaction: Building Bridges* (pp. 344–352). https://doi.org/10.1145/1463160.1463198

Shneiderman, B., Plaisant, C., Cohen, M., Jacobs, S. M., & Elmqvist, N. (2017). *Designing the user interface: Strategies for effective human-computer interaction* (6th Ed.). Pearson.

Shneiderman, B., Plaisant, C., Cohen, M., Jacobs, S., Elmqvist, N., & Diakopoulos, N. (2016). Grand challenges for HCI researchers. *Interactions, 23*(5), 24–25. https://doi.org/10.1145/2977645

Shneiderman, B., Plaisant, Cohen, Jacobs, & Elmqvist. (2018). *Designing the user interface: Strategies for effective human-computer interaction.* Pearson Education.

Shneiderman, B., Preece, J., & Pirolli, P. (2011). Realizing the value of social media requires innovative computing research. *Communications of the ACM, 54*(9), 34–37. https://doi.org/10.1145/1995376.1995389

Shore, B. (1996). *Culture in mind: Cognition, culture, and the problem of meaning.* Oxford University Press.

Siciliano, B., & Khatib, O. (Eds.). (2016). *Springer handbook of robotics.* Springer International Publishing. https://doi.org/10.1007/978-3-319-32552-1

Silva-Prietch, S., Gonzalez-Calleros, J. M., Sánchez, J. A., Olmos-Pineda, I., & Guerrero-García, J. (2019). Cultural aspects in the user experience design of an ASLR system. In *Proceedings of the IX Latin American Conference on Human Computer Interaction* (pp. 1–5). https://doi.org/10. 1145/3358961.3359000

Sim, G., Shrivastava, A., Horton, M., Agarwal, S., Haasini, P. S., Kondeti, C. S., & McKnight, L. (2019). Child-generated personas to aid design across cultures. In D. Lamas, F. Loizides, L. Nacke, H. Petrie, M. Winckler, & P. Zaphiris (Eds.), *Human-computer interaction – INTERACT 2019* (Vol. 11748, pp. 112–131). Springer International Publishing. https://doi.org/10.1007/978-3-030-29387-1_7

Singh, N., & Pereira, A. (2005). *The culturally customized web site.* Elsevier.

Smith-Jackson, T. L., Resnick, M. L., & Johnson, K. (Eds.). (2014). *Cultural ergonomics. Theory, methods, and applications.* CRC Press.

Smith, A., Bannon, L., & Gulliksen, J. (2010, March 1). Localising HCI practice for local needs. In *India HCI 2010/ Interaction Design & International Development 2010.* https://doi.org/10.14236/ ewic/IHCI2010.15

Smith, A., Dunckley, L., French, T., Minocha, S., & Chang, Y. (2012). Reprint of a process model for developing usable cross-cultural websites. *Interacting with Computers, 24*(4), 174–187. https:// doi.org/10.1016/j.intcom.2012.08.001

Smith, W., Wadley, G., Daly, O., Webb, M., Hughson, J., Hajek, J., Parker, A., Woodward-Kron, R., & Story, D. (2017). Designing an app for pregnancy care for a culturally and linguistically diverse community. In *Proceedings of the 29th Australian Conference on Computer-Human Interaction* (pp. 337–346). https://doi.org/10.1145/3152771.3152807

Smyth, M., Helgason, I., Kresin, F., Balestrini, M., Unteidig, A. B., Lawson, S., Gaved, M., Taylor, N., Auger, J., Hansen, L. K., Schuler, D. C., Woods, M., & Dourish, P. (2018). Maker movements, do-it-yourself cultures and participatory design: Implications for HCI research. In *Extended Abstracts of the 2018 CHI Conference on Human Factors in Computing Systems* (pp. 1–7). https://doi.org/10.1145/3170427.3170604

Soares, A. M., Farhangmehr, M., & Shoham, A. (2007). Hofstede's dimensions of culture in international marketing studies. *Journal of Business Research, 60*(3), 277–284. https://doi.org/10.1016/ j.jbusres.2006.10.018

Soro, A., Brereton, M., Taylor, J. L., Hong, A. L., & Roe, P. (2016). Cross-cultural dialogical probes. In *Proceedings of the First African Conference on Human Computer Interaction* (pp. 114–125). https://doi.org/10.1145/2998581.2998591

Soro, A., Brereton, M., Taylor, J. L., Lee Hong, A., & Roe, P. (2017). A cross-cultural notice-board for a remote community: Design, deployment, and evaluation. In R. Bernhaupt, G. Dalvi, A. Joshi, D. K. Balkrishan, J. O'Neill, & M. Winckler (Eds.), *Human-computer interaction— INTERACT 2017* (Vol. 10513, pp. 399–419). Springer International Publishing. https://doi.org/ 10.1007/978-3-319-67744-6_26

Spillers, F., & Gavine, A. (2018). Localization of mobile payment systems: Cultural and temporal rhythms in user adoption. In *Proceedings of the 10th Nordic Conference on Human-Computer Interaction* (pp. 862–867). https://doi.org/10.1145/3240167.3240276

Srinivasan, R., & Martinez, A. M. (2021). Cross-cultural and cultural-specific production and perception of facial expressions of emotion in the wild. *IEEE Transactions on Affective Computing, 12*(3), 707–721. https://doi.org/10.1109/TAFFC.2018.2887267

Stafford, T. F., & Duong, B. Q. (2022). Social media in emerging economies: A cross-cultural comparison. *IEEE Transactions on Computational Social Systems*, 1–19. https://doi.org/10.1109/ TCSS.2022.3169412

Stefano, M. L., & Davis, P. (Eds.). (2017). *The Routledge companion to intangible cultural heritage.* Routledge/Taylor & Francis Group.

Stojko, L. (2020). Intercultural usability of large public displays. In *Adjunct Proceedings of the 2020 ACM International Joint Conference on Pervasive and Ubiquitous Computing and Proceedings of the 2020 ACM International Symposium on Wearable Computers* (pp. 218–222). https://doi.org/10.1145/3410530.3414330

Striphas, T. (2015). Algorithmic culture. *European Journal of Cultural Studies, 18*(4–5), 395–412. https://doi.org/10.1177/1367549415577392

Sturm, C., Oh, A., Linxen, S., Abdelnour Nocera, J., Dray, S., & Reinecke, K. (2015). How WEIRD is HCI?: Extending HCI principles to other countries and cultures. In *Proceedings of the 33rd Annual ACM Conference Extended Abstracts on Human Factors in Computing Systems* (pp. 2425–2428). https://doi.org/10.1145/2702613.2702656

Suchman, L. (2011). Anthropological relocations and the limits of design. *Annual Review of Anthropology, 40*(1), 1–18. https://doi.org/10.1146/annurev.anthro.041608.105640

Suchmann, L. (1987). *Plans and situated actions: The problem of human-machine communication.* Cambridge University Press

Sun, H. (2002). Exploring cultural usability. In *Proceeding of the IEEE International Professional Communication Conference* (pp. 319–330). https://doi.org/10.1109/IPCC.2002.1049114

Sun, H. (2012). *Cross-cultural technology design: Creating culture-sensitive technology for local users.* Oxford University Press.

Sun, H. (2022). Bridging cultural differences with critical design in a globalized world. In *CHI Conference on Human Factors in Computing Systems Extended Abstracts* (pp. 1–2). https://doi.org/10.1145/3491101.3503768

Suominen, J., & Sivula, A. (2013). Gaming legacy? Four approaches to the relation between cultural heritage and digital technology. *Journal on Computing and Cultural Heritage, 6*(3), 1–18. https://doi.org/10.1145/2499931.2499933

Swidler, A. (1986). Culture in action: Symbols and strategies. *American Sociological Review, 51*(2), 273. https://doi.org/10.2307/2095521

Swidler, A. (2001a). *Talk of love: How culture matters.* University of Chicago Press.

Swidler, A. (2001b). *Talk of love.* University of Chicago Press. https://doi.org/10.7208/chicago/9780226230665.001.0001

Sylla, C., Pires Pereira, Í. S., & Sá, G. (2019). Designing manipulative tools for creative multi and cross-cultural storytelling. In *Proceedings of the 2019 on Creativity and Cognition* (pp. 396–406). https://doi.org/10.1145/3325480.3325501

Tan, P., Hills, D., Ji, Y., & Feng, K. (2020). Case study: Creating embodied interaction with learning intangible cultural heritage through WebAR. In *Extended Abstracts of the 2020 CHI Conference on Human Factors in Computing Systems* (pp. 1–6). https://doi.org/10.1145/3334480.3375199

Tang, N., & Chandra, P. (2022). Community, culture, and capital: Exploring the financial practices of older Hong Kong immigrants. In *CHI Conference on Human Factors in Computing Systems Extended Abstracts* (pp. 1–6). https://doi.org/10.1145/3491101.3519740

Taverner, J., Vivancos, E., & Botti, V. (2021). A multidimensional culturally adapted representation of emotions for affective computational simulation and recognition. *IEEE Transactions on Affective Computing,* 1–1. https://doi.org/10.1109/TAFFC.2020.3030586

Taylor, J. L., Soro, A., Brereton, M., Hong, A. L., & Roe, P. (2016a). Designing evaluation beyond evaluating design: Measuring success in cross-cultural projects. In *Proceedings of the 28th Australian Conference on Computer-Human Interaction - OzCHI'16* (pp. 472–477). https://doi.org/10.1145/3010915.3010965

Taylor, J. L., Soro, A., Hong, A. L., Roe, P., & Brereton, M. (2016b). Designing for cross-cultural perspectives of time. In *Proceedings of the First African Conference on Human Computer Interaction* (pp. 219–224). https://doi.org/10.1145/2998581.2998606

Thayer, K., Guo, P. J., & Reinecke, K. (2018). The impact of culture on learner behavior in visual debuggers. In *2018 IEEE Symposium on Visual Languages and Human-Centric Computing (VL/HCC)* (pp. 115–124). https://doi.org/10.1109/VLHCC.2018.8506556

Tibau, J., Stewart, M., Harrison, S., & Tatar, D. (2019). FamilySong: Designing to enable music for connection and culture in internationally distributed families. In *Proceedings of the 2019 on Designing Interactive Systems Conference* (pp. 785–798). https://doi.org/10.1145/3322276.3322279

Tredinnick, L. (2008). *Digital information culture: The individual and society in the digital age.* Chandos Publishing.

Triandis, H. (1995). *Individualism & collectivism.* Routledge.

Triandis, H. (2007). Culture and psychology: A history of the study of their relationship. In S. Kitayama & D. Cohen (Eds.), *Handbook of cultural psychology* (pp. 59–77). The Guildford Press.

Trompenaars, F., & Hampden-Turner, C. (2011). *Riding the waves of culture: Understanding cultural diversity in business* (2nd ed., repr. with corrections). Brealey.

Tsai, T.-W., Chang, T.-C., Chuang, M.-C., & Wang, D.-M. (2008). *Exploration in emotion and visual information uncertainty of websites in culture relations.*

Tukiainen, S. S. I. (2010). Coping with cultural dominance in cross cultural interaction. In *Proceedings of the 3rd International Conference on Intercultural Collaboration* (pp. 255–258). https://doi.org/10.1145/1841853.1841901

Tuli, A., Dalvi, S., Kumar, N., & Singh, P. (2019). "It's a girl thing": Examining challenges and opportunities around menstrual health education in India. *ACM Transactions on Computer-Human Interaction, 26*(5), 1–24. https://doi.org/10.1145/3325282

Turkle, S. (1995). *Life on the screen. Identity in the age of the internet.* Simon & Schuster.

Turkle, S. (2011). *Alone together.* Basic Books.

Tylor, E. (1871). *Primitive culture.* Cambridge University Press.

Umoja, S. (2018). *Algorithms of oppression: How search engines reinforce racism.* University Press.

Vainstein, N., Kuflik, T., & Lanir, J. (2016). Towards using mobile, head-worn displays in cultural heritage: User requirements and a research agenda. In *Proceedings of the 21st International Conference on Intelligent User Interfaces* (pp. 327–331). https://doi.org/10.1145/2856767.2856802

van der Sluis, I., Luz, S., Breitfuß, W., Ishizuka, M., & Prendinger, H. (2012). Cross-cultural assessment of automatically generated multimodal referring expressions in a virtual world. *International Journal of Human-Computer Studies, 70*(9), 611–629. https://doi.org/10.1016/j.ijhcs.2012.05.002

Van Dijck, J. (2013). *The culture of connectivity.* Oxford University Press.

van Schaik, P., Luan Wong, S., & Teo, T. (2015). Questionnaire layout and national culture in online psychometrics. *International Journal of Human-Computer Studies, 73*, 52–65. https://doi.org/10.1016/j.ijhcs.2014.08.005

Vasalou, A., Joinson, A. N., & Courvoisier, D. (2010a). Cultural differences, experience with social networks and the nature of "true commitment" in Facebook. *International Journal of Human Computer Studies, 68*(10), 719–728. https://doi.org/10.1016/j.ijhcs.2010.06.002

Vasalou, A., Joinson, A. N., & Courvoisier, D. (2010b). Cultural differences, experience with social networks and the nature of "true commitment" in Facebook. *International Journal of Human-Computer Studies, 68*(10), 719–728. https://doi.org/10.1016/j.ijhcs.2010.06.002

Vatrapu, R. K. (2010). Explaining culture: An outline of a theory of socio-technical interactions. In *Proceedings of the 3rd International Conference on Intercultural Collaboration* (pp. 111–120). https://doi.org/10.1145/1841853.1841871

Vatrapu, R., & Perez-Quinones, M. (2006). Culture and usability evaluation: The effects of culture in structured interviews. *Journal of Usability Studies, 1*(4), 156–170. http://citeseerx.ist.psu.edu/viewdoc/summary?doi=10.1.1.101.5837

Vertovec, S. (n.d.). *Towards post-multiculturalism? Changing communities, conditions and contexts of diversity.*

Visescu, I. D. (2021). The impact of culture on visual design perception. In C. Ardito, R. Lanzilotti, A. Malizia, H. Petrie, A. Piccinno, G. Desolda, & K. Inkpen (Eds.), *Human-computer interaction – INTERACT 2021* (Vol. 12936, pp. 499–503). Springer International Publishing. https://doi.org/10.1007/978-3-030-85607-6_66

Volkmar, G., Wenig, N., & Malaka, R. (2018). Memorial quest—a location-based serious game for cultural heritage preservation. In *Proceedings of the 2018 Annual Symposium on Computer-Human Interaction in Play Companion Extended Abstracts* (pp. 661–668). https://doi.org/10.1145/3270316.3271517

Wallerstein, I. (1990). Culture as the ideological battleground of the modern world-system. *Theory, Culture & Society, 7*(2–3), 31–55. https://doi.org/10.1177/026327690007002003

Walsh, T., & Nurkka, P. (n.d.). *Approaches to cross-cultural design: Two case studies with UX web-surveys.*

Walsh, T., Nurkka, P., & Walsh, R. (2010). Cultural differences in smartphone user experience evaluation. In *Proceedings of the 9th International Conference on Mobile and Ubiquitous Multimedia* (pp. 1–9). https://doi.org/10.1145/1899475.1899499

Wang, H.-C., Fussell, S. F., & Setlock, L. D. (2009). Cultural difference and adaptation of communication styles in computer-mediated group brainstorming. In *Proceedings of the SIGCHI Conference on Human Factors in Computing Systems* (pp. 669–678). https://doi.org/10.1145/1518701.1518806

Wang, I., Buchweitz, L., Smith, J., Bornholdt, L.-S., Grund, J., Ruiz, J., & Korn, O. (2020). Wow, you are terrible at this!: An intercultural study on virtual agents giving mixed feedback. In *Proceedings of the 20th ACM International Conference on Intelligent Virtual Agents* (pp. 1–8). https://doi.org/10.1145/3383652.3423887

Wang, Y. (2021). Living in a city, living a rural life: Understanding second generation mingongs' experiences with technologies in China. *ACM Transactions on Computer-Human Interaction, 28*(3), 1–29. https://doi.org/10.1145/3448979

Wang, Y., Norice, G., & Cranor, L. F. (2011). Who is concerned about what? A study of American, Chinese and Indian users' privacy concerns on social network sites. In J. M. McCune, B. Balacheff, A. Perrig, A.-R. Sadeghi, A. Sasse, & Y. Beres (Eds.), *Trust and trustworthy computing* (Vol. 6740, pp. 146–153). Springer. https://doi.org/10.1007/978-3-642-21599-5_11

Wanick, V., Stallwood, J., Ranchhod, A., & Wills, G. (2018). Can visual familiarity influence attitudes towards brands? An exploratory study of advergame design and cross-cultural consumer behaviour. *Entertainment Computing, 27*, 194–208. https://doi.org/10.1016/j.entcom.2018.07.002

Washington, A. N. (2020). When twice as good isn't enough: The case for cultural competence in computing. In *Proceedings of the 51st ACM Technical Symposium on Computer Science Education* (pp. 213–219). https://doi.org/10.1145/3328778.3366792

Weber, F., Chadowitz, R., Schmidt, K., Messerschmidt, J., & Fuest, T. (2019). Crossing the street across the globe: A study on the effects of eHMI on pedestrians in the US, Germany and China. In H. Krömker (Ed.), *HCI in mobility, transport, and automotive systems* (Vol. 11596, pp. 515–530). Springer International Publishing. https://doi.org/10.1007/978-3-030-22666-4_37

Wecker, A. J., Kuflik, T., & Stock, O. (2017). AMuse: Connecting indoor and outdoor cultural heritage experiences. In *Proceedings of the 22nd International Conference on Intelligent User Interfaces Companion* (pp. 153–156). https://doi.org/10.1145/3030024.3040980

Welzer, T., Hölbl, M., Družovec, M., & Brumen, B. (2011). Cultural awareness in social media. In *Proceedings of the 2011 International Workshop on DETecting and Exploiting Cultural DiversiTy on the Social Web* (pp. 3–8). https://doi.org/10.1145/2064448.2064463

Williams, M. J., & Spencer-Rodgers, J. (2010). Culture and stereotyping processes: Integration and new directions: Culture and stereotyping. *Social and Personality Psychology Compass, 4*(8), 591–604. https://doi.org/10.1111/j.1751-9004.2010.00288.x

Winograd, T. (1997). *From computing machinery to interaction design.*

Winschiers-Theophilus, H., Chivuno-Kuria, S., Kapuire, G. K., Bidwell, N. J., & Blake, E. (2010). Being participated: A community approach. In *Proceedings of the 11th Biennial Participatory Design Conference* (pp. 1–10). https://doi.org/10.1145/1900441.1900443

Winschiers, D. H. (n.d.). *The challenges of participatory design in an intercultural context: Designing for usability in Namibia.*

Winschiers, H. (2006). The challenges of participatory design in an intercultural context: Designing for usability in Namibia. In *Proceedings of the Participatory Design Conference* (pp. 73–76). http://ojs.ruc.dk/index.php/pdc/article/view/375

Winschiers, H., & Fendler, J. (2007). Assumptions considered harmful the need to redefine usability. *Usability and Internationalization*, Part I, HCII 2007 Ed N. Aykin.

Wong-Villacres, M., DiSalvo, C., Kumar, N., & DiSalvo, B. (2020). Culture in action: Unpacking capacities to inform assets-based design. In *Proceedings of the 2020 CHI Conference on Human Factors in Computing Systems* (pp. 1–14). https://doi.org/10.1145/3313831.3376329

Wu, H., Gai, J., Wang, Y., Liu, J., Qiu, J., Wang, J., & Zhang, X. (2020). Influence of cultural factors on freehand gesture design. *International Journal of Human-Computer Studies, 143*, 102502. https://doi.org/10.1016/j.ijhcs.2020.102502

Wyche, S. (2019). Using cultural probes in new contexts: Exploring the benefits of probes in HCI4D/ICTD. In *Conference Companion Publication of the 2019 on Computer Supported Cooperative Work and Social Computing* (pp. 423–427). https://doi.org/10.1145/3311957.3359454

Xie, B., & Jaeger, P. T. (2008). Older adults and political participation on the internet: A cross-cultural comparison of the USA and China. *Journal of Cross-Cultural Gerontology, 23*(1), 1–15. https://doi.org/10.1007/s10823-007-9050-6

Yakura, H. (2021). No more handshaking: How have COVID-19 pushed the expansion of computer-mediated communication in Japanese idol culture? In *Proceedings of the 2021 CHI Conference on Human Factors in Computing Systems* (pp. 1–10). https://doi.org/10.1145/3411764.3445252

Yalamu, P., Chua, C., & Doube, W. (2019). Does indigenous culture affect one's view of an LMS interface: A PNG and Pacific Islands students' perspective. In *Proceedings of the 31st Australian Conference on Human-Computer-Interaction* (pp. 302–306). https://doi.org/10.1145/3369457.3369483

Yammiyavar, P., Clemmensen, T., & Kumar, J. (2008). *Influence of cultural background on non-verbal communication in a usability testing situation.*

You, S., Kim, M., & Lim, Y. (2016). Value of culturally oriented information design. *Universal Access in the Information Society, 15*(3), 369–391. https://doi.org/10.1007/s10209-014-0393-9

Zaharias, P., & Papargyris, A. (2009). The gamer experience: Investigating relationships between culture and usability in massively multiplayer online games. *Computers in Entertainment, 7*(2), 1–24. https://doi.org/10.1145/1541895.1541906

Zahed, F., Van Pelt, W. V., & Song, J. (2001). A conceptual framework for international web design. *IEEE Transactions on Professional Communication, 44*(2), 83–103. https://doi.org/10.1109/47. 925509

Zamora, J. (2019). Are we having fun yet?: Designing for fun in artificial intelligence that is multi-cultural and multiplatform. In *Proceedings of the 7th International Conference on Human-Agent Interaction* (pp. 208–210). https://doi.org/10.1145/3349537.3352767

Zhan, J., Liu, M., Garrod, O. G. B., Jack, R. E., & Schyns, P. G. (2020). A generative model of cultural face attractiveness. In *Proceedings of the 20th ACM International Conference on Intelligent Virtual Agents* (pp. 1–3). https://doi.org/10.1145/3383652.3423914

Zhang, Y., Zong, R., Kou, Z., Shang, L., & Wang, D. (2022). CollabLearn: An uncertainty-aware crowd-AI collaboration system for cultural heritage damage assessment. *IEEE Transactions on Computational Social Systems, 9*(5), 1515–1529. https://doi.org/10.1109/TCSS.2021.3109143

Zhao, C., & Jiang, G. (2011). Cultural differences on visual self-presentation through social networking site profile images. In *Proceedings of the SIGCHI Conference on Human Factors in Computing Systems* (pp. 1129–1132). https://doi.org/10.1145/1978942.1979110

Zhao, C., Hinds, P., & Gao, G. (2012). How and to whom people share: The role of culture in self-disclosure in online communities. In *Proceedings of the ACM 2012 Conference on Computer Supported Cooperative Work* (pp. 67–76). https://doi.org/10.1145/2145204.2145219

Zheng, C., Khan, A. H., & Matthews, B. (2018). Bridging the cross-cultural language divide through design. In *Proceedings of the 30th Australian Conference on Computer-Human Interaction* (pp. 167–171). https://doi.org/10.1145/3292147.3292222

Zhou, L., Sun, X., Mu, G., Wu, J., Zhou, J., Wu, Q., Zhang, Y., Xi, Y., Gunes, N. D., & Song, S. (2022). A tool to facilitate the cross-cultural design process using deep learning. *IEEE Transactions on Human-Machine Systems, 52*(3), 445–457. https://doi.org/10.1109/THMS.2021.312 6699

Zhou, S., Zhang, Z., & Bickmore, T. (2017). Adapting a persuasive conversational agent for the Chinese culture. In *Proceedings - 2017 International Conference on Culture and Computing, Culture and Computing 2017*, 2017-December (pp. 89–96). https://doi.org/10.1109/Culture.and.Comput ing.2017.42

Printed in the United States
by Baker & Taylor Publisher Services